ENGLISH FOR
CUISINE
Second Edition

烹饪英语

（第2版）

主　编　张　媛
副主编　胡　音　郑元珂
参　编　王　畅　李潇潇
　　　　黄　巧　李　杨
　　　　彭施龙

重庆大学出版社

图书在版编目（CIP）数据

烹饪英语 / 张媛主编. — 2版. — 重庆: 重庆
大学出版社, 2023.11

ISBN 978-7-5689-3820-4

I. ①烹… II.①张… III.①烹饪－英语 IV.
①TS972.1

中国国家版本馆CIP数据核字（2023）第052427号

烹饪英语（第2版）

张 媛 主编

责任编辑：陈 亮 版式设计：陈 亮
责任校对：邹 忌 责任印制：赵 晟

*

重庆大学出版社出版发行

出版人：陈晓阳

社址：重庆市沙坪坝区大学城西路21号

邮编：401331

电话：（023）88617190 88617185（中小学）

传真：（023）88617186 88617166

网址：http://www.cqup.com.cn

邮箱：fxk@cqup.com.cn（营销中心）

全国新华书店经销

重庆天旭印务有限责任公司印刷

*

开本：787mm×1092mm 1/16 印张：10.25 字数：301千
2021年12月第1版 2023年11月第2版 2023年11月第2次印刷
ISBN 978-7-5689-3820-4 定价：39.00元

前　言

进入新时代，中国要向世界阐释推介更多具有中国特色、体现中国精神、蕴藏中国智慧的优秀文化，以展示真实、立体、全面的中国，形成同我国综合国力和国际地位相匹配的国际话语权。

饮食文化是国家软实力的重要组成部分。据中国外文局历年发布的中国国家形象全球调查报告，在海外受访者眼中，饮食一直都是体现中国文化的代表性元素。将饮食文化作为切入点讲述中国故事，以文化人，能更好地展示中国形象，体现中华魅力。然而，中西文化差异造成的文化"失语"现象一直是个困扰饮食文化讲述者的主要问题。

本教材着力培养学习者用英语讲述中国饮食文化的能力，围绕饮食概述、节庆饮食、餐饮礼仪、厨房用具、食材调料、调味及烹饪方法等内容，以文本输入的方式引导学习者学习语言的同时，挖掘中国饮食文化，强调基于内容的语言教学。在各章节每篇阅读文章之后，设置讨论、写作、翻译、短视频制作等输出任务，让学习者进一步发现身边的饮食文化，进行探究性学习，培养其文化自觉和文化自信，锻炼其用英语讲述中国饮食文化的能力。

本教材的出版既是团队成员辛苦付出的结果，也离不开重庆大学出版社的大力支持，在此表示感谢。

因编者水平有限，教材内容难免有疏漏之处，请各位读者不吝赐教，以便之后修订完善。

编　者

2023年5月

CONTENTS

CHAPTER V INGREDIENTS

CHAPTER VI SEASONINGS AND SPICES

CHAPTER VII FLAVORING

CHAPTER VIII COOKING METHODS

CHAPTER IX NOTABLE FOODS

GLOSSARY

CHAPTER I

PROFILE

Chinese Cuisine

The old Chinese saying that food is the **paramount** necessity of people fully exhibits the significance of diet for Chinese people. Chinese cuisine, an **inclusive** term indicating the various foods from across China, is an important part of Chinese culture. In fact, it has been considered to best **represent** China along with Chinese medicine and martial arts, according to **China National Image Global Survey**.

Chinese cuisine boasts a wide variety due to variations in geography, climate, **produce**, culture and even beliefs. It is usually **categorized** into four or eight main schools referring to their origins. The four most renowned traditions are Lu cuisine of the lower reaches of the Yellow River, Sichuan cuisine of the upper reaches of the Yangtze River, Huaiyang cuisine of Huai'an and Yangzhou in Jiangsu Province, and Cantonese cuisine of the Lingnan region. Dishes are usually grouped into cold dishes, hot dishes, main dishes, snacks and soups.

Taste and smell are not the only things that Chinese cuisine emphasizes. Quality of a dish is usually appraised or described from six aspects, namely, color, smell, taste, appearance, meaning and health value. An ideal dish is not only supposed to appeal to the eye, nose and **palate**, but convey positive connotations and benefit body health as well. For example, Sixi Wanzi, or Braised Pork Balls in Gravy, is a reflection of people's good wishes for a happy life. The spicy hot pot of Sichuan helps the body to **excrete** extra water, remove inner humidity and maintain body balance.

Tea, enjoyed by people from all social classes, plays an important role in Chinese dining culture. It is consumed throughout the day, including during meals, as a substitute for plain water, or for simple pleasure. In restaurants in China, tea is usually served **in lieu of** water. When sipped as a daily beverage, Chinese people tend to use a special personal tea bottle, in which water is allowed to **infuse** with tea leaves for hours. Chinese tea is often classified into several different categories according to the species of plant from which it is sourced, the region in which it is grown, and the method of production used. The different types of Chinese tea include black, white, green, yellow, oolong and dark tea.

Alcoholic beverages such as Baijiu and Huangjiu are preferred by many adults as part of diet and ceremonies. They were traditionally warmed before being consumed, and the temperature to which the liquor may be warmed ranges between approximately 35 °C and 55 °C, well below the boiling point of ethanol. Warming the liquor allows its **aromas** to be better appreciated by the drinker without losing too much alcohol. The optimal temperature for warming depends on the type of beverage as well as the preference of the drinker.

Chinese culinary culture has influenced many other cuisines in Asia, with modifications made to cater to local palates, and **utensils** such as chopsticks and the **wok** can now be found worldwide.

Vocabulary

paramount /ˈpærəmaʊnt/ *adj.* 至关重要的
inclusive /ɪnˈkluːsɪv/ *adj.* 范围广泛的
represent /reprɪˈzent/ *v.* 代表
China National Image Global Survey 中国国家形象调查报告

produce /ˈprɒdjuːs/ *n.* 物产
categorize /ˈkætəgəraɪz/ *v.* 分类
palate /ˈpælət/ *n.* 味觉
excrete /ɪkˈskriːt/ *v.* 排出
in lieu of 替代

infuse /ɪnˈfjuːz/ v. 注入

utensil /juːˈtensl/ n. 用具

aroma /əˈrəʊmə/ n. 香味

wok /wɒk/ n. 炒锅

Tasks

 I. Group Talk

Foods play a very important role in Chinese people's life, and words describing foods and cookery are often used to convey meanings beyond the culinary field. For example, "Chicu", literally "drinking vinegar", means jealousy. Can you think of more examples? Talk with your partners and share your opinions.

 II. Writing

Yi Yin, a famous politician in the Shang Dynasty, is considered the ancestor of Chinese chefs. Gifted in cooking, he compared state administration to cookery, and developed the famous philosophy of "balance of five tastes". Learn more about his theory and write an article of about 80 words.

_____

_____

_____

_____

_____

_____

_____

_____

Ⓔ III. Translation

Translate the following sentences into English.

1. 中国菜讲究色、香、味、形、意、养。

   _____

   _____

2. 月饼又称团圆饼，是中国传统美食之一，也是中秋节必备美食。

   _____

   _____

3. 一年四季，按季节而吃，是中国烹饪的一大特征。

   _____

   _____

4. 菜肴多数根据主、辅、调料及烹调方法进行命名，少数根据历史掌故、神话传说、菜肴形象等来命名。

   _____

   _____

5. 中国人认为"医食同源""药膳同功"，所以会利用食物原料的药用价值，做成各种养生佳肴。

   _____

   _____

▶ IV. Short Video

Make a short video about the cuisine of your hometown, and share it with your classmates.

Sichuan Cuisine

Passage Ⓑ

While China is an ideal place to feed your **appetite**, Sichuan is the best spot to satisfy your palate.[1] Sichuan cuisine, one of the most renowned regional flavors originating in the southwest of China, is well received by an increasing number of people around China and even around the world. It is reputed for its wide range of **ingredients**, varied ways of seasoning and **diverse** cooking techniques. The result is a cuisine with an incredible depth and **complexity** of flavors, hitting all **sense receptors** in your mouth, nose, and **gastrointestinal** system at the same time. UNESCO declared Chengdu, the capital of Sichuan Province, to be a city of **gastronomy** in 2011 to recognize the **sophistication** of its cooking.

There have been six major migration waves in the history of Sichuan. People from different regions and ethnic groups had brought with them their **distinct** dietary habits and cooking techniques. These habits and techniques **merged** and interacted with local cuisine, contributing to the **acclaimed** diversity of Sichuan cookery and the incredible creation of Sichuan chefs. It is widely agreed that the best cuisines in the world are those with an **enthusiasm** for and ability to **incorporate** ingredients from around the globe. Chilies are not a native produce, but it is hard to imagine Sichuan cooking without chilies nowadays. As Fuchsia Dunlop writes in *Sichuan Cooking*, "Local cooks are beginning to experiment with Japanese **wasabi**, all kinds of fresh seafood and Korean-influenced barbecues." She continues, "Few local chefs or **gourmets** see this as a threat: such is their confidence in Sichuanese cooking skills that they are merely excited at the **prospect** of having so many new ingredients and techniques to play with."

Sichuan cooking has **bold** flavors, particularly the **pungency** and spiciness resulting from **liberal** use of garlic, spring onions, ginger and chili peppers, as well as the unique flavor of **Sichuan pepper**. In Sichuan Basin, heavy, humid skies create a wet, grey blanket for most of the year, slowing down the body. Spicy and pungent foods are believed to be perfect for the Sichuan climate, for they help the body to excrete water, whether during cold humid winters or hot humid summers.

One of the best known Sichuan foods is hot pot. Hot pot is a cooking method, prepared with a **simmering** pot of soup stock at the dining table. While a hot pot full of flavored **broth** is kept simmering, raw ingredients, which are usually pre-sliced into thin sections, are placed into the pot and are cooked. Most raw foods can be cooked in a hot pot, although they may have different cooking times, and must be **immersed** in the soup and then removed accordingly. The cooked food is often eaten with a dipping sauce for additional flavoring. The dipping sauce is usually self-made by diners on the spot with mashed garlic, sesame oil, chopped spring onions and coriander, salt and oyster oil, etc.

Although renowned for spiciness, Sichuan cuisine features a lot of **delicate** and savory dishes, and a typical Sichuanese meal includes non-spicy dishes to cool and **soothe** the palate. Just as the saying goes, Sichuan cooking is of "one dish, one shape; hundreds of dishes, hundreds of tastes"[2].

Notes

[1] While China is an ideal place to feed your appetite, Sichuan is the best spot to satisfy your palate. 食在中国，味在四川。

[2] one dish, one shape; hundreds of dishes, hundreds of tastes 一菜一格，百菜百味

Vocabulary

appetite /ˈæpɪtaɪt/ *n.* 胃口

ingredient /ɪnˈɡriːdiənt/ *n.* 食材

diverse /daɪˈvɜːs/ *adj.* 多样的

complexity /kəmˈpleksəti/ *n.* 复杂性

sense receptor 感官

gastrointestinal /ɡæstrəʊɪnˈtestɪnl/ *adj.* 肠胃的

gastronomy /ɡæˈstrɒnəmi/ *n.* 美食

sophistication /səˌfɪstɪˈkeɪʃn/ *n.* 成熟

distinct /dɪˈstɪŋkt/ *adj.* 不同的

merge /mɜːdʒ/ *v.* 融入

acclaim /əˈkleɪm/ *v.* 称赞

enthusiasm /ɪnˈθjuːziæzəm/ *n.* 热情

incorporate /ɪnˈkɔːpəreɪt/ *v.* 合并

wasabi /wəˈsɑːbi/ *n.* 绿芥末

gourmet /ˈɡʊəmeɪ/ *n.* 美食家

prospect /ˈprɒspekt/ *n.* 前景；前途

bold /bəʊld/ *adj.* 大胆的

pungency /ˈpʌndʒənsi/ *n.* 辛辣

liberal /ˈlɪbərəl/ *adj.* 慷慨的

Sichuan pepper 花椒

simmer /ˈsɪmə(r)/ *v.* 文火炖

broth /brɒθ/ *n.* 浓汤

immerse /ɪˈmɜːs/ *v.* 沉浸

delicate /ˈdelɪkət/ *adj.* 美味的

soothe /suːð/ *v.* 抚慰

Tasks

 I. Group Talk

Besides pungency and spiciness, are there any other Sichuan flavors that strike you most? Talk with your partners and share your opinions.

 II. Writing

Sichuan snacks are becoming increasingly popular among tourists. Do your research and write an article of about 80 words to introduce your favorite snacks.

_____

_____

_____

_____

_____

_____

III. Translation

Translate the following sentences into English.

1. 清朝中期，由于运用辣椒调味，川菜有了进一步发展。

   _____

   _____

2. 四川优越的自然条件为川菜提供了丰富而优质的烹饪原料。

   _____

   _____

3. 川菜取材广泛，但不以食材的古怪和稀缺为号召力，而是以普通、绿色、健康为选材的基本原则。

   _____

   _____

4. 外地人谈到川菜，常常认为川菜的风味特点就只是麻辣，其实这个看法失之偏颇。

   _____

   _____

5. 川菜真正的风味特色是清、鲜、醇、浓并重，善用麻辣。

   _____

   _____

IV. Short Video

Make a short video about an ingredient, spice or seasoning characteristic of Sichuan, and share it with your classmates and friends.

Chinese Mianshi[1]

Mianshi, flour food if translated literally, is a Chinese culinary term referring to the various foods that use wheat as the main ingredient, such as Mantou, Baozi, Jiaozi and noodles, etc.

Mantou[2], often known as Chinese bread, is a type of white steamed **bun fluffy** in **texture** and popular in the northern parts of China where it is sometimes called Mo instead of Mantou. The buns are made with **milled** wheat flour, water and **leavening agents**, and eaten as a **staple** food in northern Chinese diet, **analogous** to rice, which forms the **mainstay** of the southern culinary tradition. Mantou can be further **manipulated**, mostly in restaurants, by deep frying and dipping in sweetened **condensed milk**. Colors and flavors may be added with other ingredients from brown sugar to food coloring, often reserved for special occasions, when they are even kneaded into various shapes in Shanxi, Shaanxi, and Shandong called Huamo[3] or colorful Mantou, a national **intangible** cultural heritage property in China.

Baozi or Bao refers to filled buns in various Chinese cuisines. There are many variations in **fillings** (meat or vegetarian) and preparations, though the buns are most often steamed. Two types are found in most parts of China: Dabao (big bun), measuring about ten centimeters across, and Xiaolongbao (small bun), measuring approximately five centimeters wide. In restaurants, a small ceramic dish for dipping the Baozi is provided with vinegar or soy sauce, both of which are available in bottles at the table, along with various types of chili and garlic pastes.

Jiaozi or Chinese dumpling is one of the major dishes eaten during the Chinese New Year. The dumplings are folded to resemble Chinese **sycee** and have great cultural significance attached to them. They typically consist of a filling wrapped into a thinly rolled piece of dough, which is then **sealed** by pressing the edges together. Common meat fillings include pork, beef, lamb and shrimps, which are usually mixed with chopped vegetables like **napa cabbage**, **scallions**, **garlic chives**, celery, lotus roots and mushrooms. Finished Jiaozi can be boiled, steamed or fried, and served with vinegar or in a soup. Though considered part of Chinese cuisine, Jiaozi is popular in other parts of East Asia and in the Western world, where a fried variety is sometimes called **potstickers** in North America and Chinese dumplings in the UK and Canada.

Noodles were invented in China, and are an essential ingredient and staple in Chinese cuisine. They are usually made from unleavened wheat dough which is either rolled flat and cut, stretched, or **extruded**, into long **strips** or **strings**. Noodles are usually cooked in boiling water, sometimes with cooking oil or salt added, and often served with an accompanying sauce or in a soup. They can be dried and stored for future use. There are over 1,200 types of noodles commonly consumed in China today, varying in size and shape. Chinese noodles have also entered the cuisines of neighboring East Asian countries such as the Democratic People's Republic of Korea and Japan, as well as Southeast Asian countries such as Vietnam, Cambodia, and Thailand.

Notes

[1] Mianshi 面食，主要指以面粉制作而成的食物。

[2] Mantou 馒头，北方部分地区也叫馍。

[3] Huamo 花馍，也叫面花，民间面塑品，2008 年被列为国家级非物质文化遗产。山西省运城市的闻喜花馍因花式各样而闻名全国，已有 1000 多年历史，形成了独特的艺术风格和完整的创作体系。

Vocabulary

bun /bʌn/ n. 小圆面包

fluffy /ˈflʌfi/ adj. 松软的

texture /ˈtekstʃə(r)/ n. 质地；口感

mill /mɪl/ v. 磨成粉

leavening agent 发酵剂

staple /ˈsteɪpl/ adj. 主要的

analogous /əˈnæləgəs/ *adj.* 相似的

mainstay /ˈmeɪnsteɪ/ *n.* 支柱

manipulate /məˈnɪpjuleɪt/ *v.* 操作

condensed milk 炼乳

intangible /ɪnˈtændʒəbl/ *adj.* 无形的；非
物质的

filling /ˈfɪlɪŋ/ *n.* 馅；填充

sycee /saɪˈsiː/ *n.* 元宝；银锭

seal /siːl/ *v.* 密封

napa cabbage 大白菜

scallion /ˈskæliən/ *n.* 葱

garlic chive 韭菜

potsticker /ˈpɒtstɪkə/ *n.* 锅贴

extrude /ɪkˈstruːd/ *v.* 挤压

strip /strɪp/ *n.* 条

string /strɪŋ/ *n.* 线

Tasks

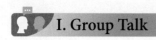

I. Group Talk

Have you tried Italian pasta? Do you know its similarities with and differences from Chinese noodles? Do your research and share your opinions with your classmates.

II. Writing

What's your favorite Mianshi? Write an article of about 80 words to introduce it.

III. Translation

Translate the following sentences into English.

1. 山西面食是我国传统特色面食文化的代表，被称为"世界面食之根"。

   _____

   _____

2. 山西刀削面、北京炸酱面、河南烩面、湖北热干面、四川担担面，合称中国五大面食。

   _____

   _____

3. 北方民间至今还流传着"冬至不端饺子碗，冻掉耳朵没人管"的说法。

   _____

   _____

4. 在中国北方地区，到了大年三十的晚上，最重要的活动就是全家老小一起包饺子。

   _____

   _____

5. 四川有许多不起眼的小馆子，卖的是让人垂涎三尺的面条。

   _____

   _____

IV. Short Video

Make a short video on how to make Jiaozi and share it with your classmates and friends.

Supplementary Reading

World Cuisines

There are three most renowned culinary regions across the globe: Eastern, Arabian and Western.

Eastern cuisines refer to those popular in eastern, northeastern and southeastern areas of Asia, including China, Japan, the Democratic People's Republic of Korea, Thailand, Singapore and Myanmar. Chopsticks are the typical eating utensils in these areas. Foods, considered remedies for health, are usually grouped into staples and non-staples. Rice and wheat are the most common staple foods, and pork is the most consumed meat.

Arabian or Islamic cookery, born on the Arabian Peninsula, is the various regional cuisines spanning Western Asia, Southern Asia, Middle Africa and Northern Africa. It is generally characterized by fragrant and copious spices, nuts, olive oil, and creamy elements. Lamb and mutton is the predominant meat as a result of religious laws banning pork, and chicken, camel, beef and fish are also used, but less frequently. In these traditions, it is common for diners to take their food from a communal plate in the center of the table. They traditionally do not use forks or spoons; instead they scoop up the food with a pita or a thumb and two fingers.

Western cooking, the culinary traditions in Europe, Americas and Oceania, derives its base from French cuisine, and the knife and fork is the most common eating utensil used at table. From its early French beginnings, Western cuisine began to expand by accepting the influences of various European countries, especially Italy and Spain. This introduced a new world of dishes such as pasta, which has since become one of the main components of Western cuisine. Despite its ever-evolving definition, Western food does have certain rules that stay true to the classics. It is still common to see a plate of protein (either meat, fish or poultry) in combination with a starchy side (potatoes, rice or pasta) and some vegetables for a main course. Soups and salads are still served as starters, and the meal still closes with a dessert at the end. Many dairy products are utilized in cooking, and there are hundreds of varieties of cheese and other fermented milk products.

Recipe Sharing

Gongbao Chicken

Ingredients

200g chicken; 40g crispy peanuts; 10g dried chilies, cut into sections; 4g Sichuan pepper; 8g ginger, sliced; 10g garlic, sliced; 15g scallion, finely chopped; 60g cooking oil

Seasonings A

0.5g salt, 5g Shaoxing cooking wine, 3g soy sauce, 10g cornstarch-water mixture

Seasonings B

1g salt, 5g Shaoxing cooking wine, 7g soy sauce, 10g vinegar, 10g sugar, 1g MSG, 15g cornstarch batter, 20g stock

Preparation

1. Cut the chicken into 1.5cm cubes, add Seasonings A and mix well.

2. Mix Seasonings B to make the thickening sauce.

3. Heat oil in a wok to 140℃, add the dried chilies and Sichuan pepper, and stir-fry to bring out the aroma. Add the diced chicken, stir-fry till just cooked, and then add ginger, garlic, scallion and the thickening sauce. Add the crispy peanuts when the sauce is thick and lustrous, mix evenly and then transfer to a serving dish.

宫保鸡丁

食材配方

鸡肉 200 克、酥花生 40 克、干辣椒节 10 克、花椒 4 克、姜片 8 克、蒜片 10 克、葱丁 15 克、食用油 60 克

调料 A： 食盐 0.5 克、料酒 5 克、酱油 3 克、水淀粉 10 克

调料 B： 食盐 1 克、料酒 5 克、酱油 7 克、醋 10 克、白糖 10 克、味精 1 克、水淀粉 15 克、鲜汤 20 克

制作工艺

1. 鸡肉斩成 1.5 厘米的丁，加入调料 A 拌匀。

2. 调料 B 混合后装入调味碗成芡汁。

3. 锅中烧油至 140℃，放入干辣椒、花椒炒香，放入鸡丁炒至断生，加入姜片、蒜片、葱丁炒香，倒入芡汁，待收汁亮油时放入酥花生炒匀，装盘。

Recipe Sharing

Tom Yum Kung

Ingredients

50g tom yum paste, 10g galangal, 2 lemon leaves, 10g lemongrass, 3g fish sauce, 150g prawns, 30g baby squids, 150g blue crabs, 30g straw mushrooms, 30g tomatoes, 3g large coriander, 3g basil, 150g coconut cream, 30g salad oil

Preparation

1. Primary processing of seafood: Peel, clean and devein prawns; clean baby squids and cut them into rings; pre-process and boil blue crabs, set aside the fleshy parts for future use.

2. Primary processing of vegetables: Boil straw mushrooms and cut each into halves; wedge tomatoes; slice galangal and lemongrass.

3. Soup base preparation: Add salad oil to the wok, fry prawn heads and blue crabs' offal; add galangal, lemongrass, fish sauce, tom yum paste, lemon leaves, water and coconut cream, and stew for 15 minutes. Sieve the mixture to get the soup base.

4. Cooking: Boil various seafood and vegetables in the sieved soup base. For plating, place the seafood and vegetables in the plate before pouring in soup; sprinkle with chopped coriander and basil.

冬阴功汤

食材配方

冬阴功酱 50 克、南姜 10 克、柠檬叶两片、香茅草 10 克、鱼露 3 克、大虾 150 克、小鱿鱼仔 30 克、蓝花蟹 150 克、草菇 30 克、番茄 30 克、大叶香菜 3 克、九层塔 3 克、椰浆 150 克、色拉油 30 克

制作工艺

1. 初加工各种海鲜：大虾去壳、去沙线，开背后清洗干净备用；小鱿鱼清洗干净切圈备用；蓝花蟹初加工后煮熟，取肉多的部分备用。

2. 初加工蔬菜：草菇汆水去掉涩味，对开备用；番茄切角备用；南姜、香茅草切片备用。

3. 加工汤底：锅内放入色拉油炒香大虾头、蓝花蟹下脚料，加入南姜、香茅草、鱼露、冬阴功酱、柠檬叶、清水、大量椰浆，调制好味道咸淡即可，熬制 15 分钟即可过滤备用。

4. 制作：过滤后的冬阴功汤煮熟蔬菜和海鲜即可装盘。先放入各种海鲜和蔬菜，然后注入汤汁，撒大叶香菜碎、九层塔碎即可成菜。

CHAPTER II   FOOD FOR SPECIAL OCCASIONS

Twenty-Four Solar Terms and Eating Customs

Passage Ⓐ

The twenty-four solar terms are knowledge of time and **practices** developed in China through observation of the sun's annual motion, according to UNESCO[1]. The ancient Chinese divided the sun's annual circular motion into twenty-four **segments**, calling each segment a "solar term". The criteria for their **formulation** were developed through the observation of changes of seasons, **astronomy** and other natural **phenomena**. The terms remain important to farmers for guiding their practices, and some **rituals** and **festivities** are also associated with the terms.

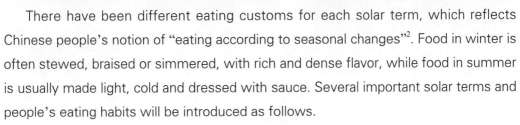

There have been different eating customs for each solar term, which reflects Chinese people's notion of "eating according to seasonal changes"[2]. Food in winter is often stewed, braised or simmered, with rich and dense flavor, while food in summer is usually made light, cold and dressed with sauce. Several important solar terms and people's eating habits will be introduced as follows.

The Beginning of Spring[3] is the first solar term among the twenty-four solar terms. People in many places make traditional food like spring rolls or spring pancakes, which are also called "biting the spring", meaning taking a bite of what the spring can offer. People may also enjoy some fresh vegetables in spring to **prevent** disease and also to welcome the new spring.

When it comes to the Beginning of Summer, duck eggs are considered by the **traditional Chinese medicine** as the first choice of food to **supplement** iron and calcium in the body and **replenish** one's energy during summer days. Also, eating duck eggs expresses people's wish for everything to be smooth, as duck eggs are round, complete and smooth. Drinking mung bean soup is also recommended, for it can **relieve** the summer heat.

Lesser Fullness of Grain falls in a period of the temporary shortage when last season's crops are almost eaten up and the new ones are not yet ripe. Therefore, people in the past had to eat bitter herbs, such as sow thistle. People also believe that the hot weather in May would cause harmful effects on the body, so eating something bitter to relieve internal poison in body will help.

The Winter **Solstice** marks the incoming of the coldest days in winter. Cold weather would block blood **circulation** and make people's ears reddened from the cold. Northern people usually eat dumplings to **protect** themselves **against** the cold, since the shape of dumplings is very similar to that of people's ears and they believe that eating dumplings will protect their ears from **freezing off**. Southern people often eat meat, especially lamb, and drink lamb soup to increase heat in body and keep warm. People in southern China also eat Tangyuan (also called **sweet dumplings**) on this day, yet they may also eat dumplings as they like.

Greater Cold is the last among the twenty-four solar terms. People tend to enjoy sticky rice, which can nourish the stomach and defend people from the cold. The most typical food made of sticky rice is eight-treasure rice pudding, that is, steamed sticky rice mixed with lard, bean paste, red dates, Job's tears, lotus seeds, dried longan pulp and preserved fruits.

It's obvious that the twenty-four solar terms have exerted great influences on human being's **biorhythms**, habits and eating customs. Living, working and eating according to the terms and different seasons are in accordance with Chinese people's philosophical thinking that man is an integral part of nature.

Notes

[1] UNESCO (United Nations Educational, Scientific and Cultural Organization) 联合国教科文组织

[2] eating according to seasonal changes 应季而食

³ 二十四节气中英文对照（来自中国气象局）：

立春 the Beginning of Spring（1st solar term）Feb. 3, 4, or 5

雨水 Rain Water（2nd solar term）Feb. 18, 19 or 20

惊蛰 the Waking of Insects（3rd solar term）Mar. 5, 6, or 7

春分 the Spring Equinox（4th solar term）Mar. 20, 21 or 22

清明 Pure Brightness（5th solar term）Apr. 4, 5 or 6

谷雨 Grain Rain（6th solar term）Apr. 19, 20 or 21

立夏 the Beginning of Summer（7th solar term）May 5, 6 or 7

小满 Lesser Fullness of Grain（8th solar term）May 20, 21 or 22

芒种 Grain in Beard（9th solar term）Jun. 5, 6 or 7

夏至 the Summer Solstice（10th solar term）Jun. 21 or 22

小暑 Lesser Heat（11th solar term）Jul. 6, 7 or 8

大暑 Greater Heat（12th solar term）Jul. 22, 23 or 24

立秋 the Beginning of Autumn（13th solar term）Aug. 7, 8 or 9

处暑 the End of Heat（14th solar term）Aug. 22, 23 or 24

白露 White Dew（15th solar term）Sep. 7, 8 or 9

秋分 the Autumn Equinox（16th solar term）Sep. 22, 23 or 24

寒露 Cold Dew（17th solar term）Oct. 8 or 9

霜降 Frost's Descent（18th solar term）Oct. 23 or 24

立冬 the Beginning of Winter（19th solar term）Nov. 7 or 8

小雪 Lesser Snow（20th solar term）Nov. 22 or 23

大雪 Greater Snow（21st solar term）Dec. 6, 7 or 8

冬至 the Winter Solstice（22nd solar term）Dec. 21, 22 or 23

小寒 Lesser Cold（23rd solar term）Jan. 5, 6 or 7

大寒 Greater Cold（24th solar term）Jan. 20 or 21

Vocabulary

practice /ˈpræktɪs/ n. 做法；活动

segment /ˈsegmənt/ n. 部分

formulation /ˌfɔːmjuˈleɪʃn/ n. 形成

astronomy /əˈstrɒnəmi/ n. 天文学

phenomenon /fɪˈnɒmɪnən/ n. 现象（复数为 phenomena）

ritual /ˈrɪtʃuəl/ n. 仪式；礼制

festivity /feˈstɪvəti/ n. 欢庆；庆典

prevent /prɪˈvent/ v. 妨碍；阻止

traditional Chinese medicine 传统中药

supplement /ˈsʌplɪmənt/ v. 增补；补充

replenish /rɪˈplenɪʃ/ v. 再度装满

relieve /rɪˈliːv/ v. 解除；减轻

solstice /ˈsɒlstɪs/ n. 至；至日；至点

circulation /ˌsɜːkjəˈleɪʃn/ n. 循环；流动

protect... against... 保护……使免受……

freeze off 冻坏

sweet dumpling 汤圆

biorhythm /ˈbaɪəʊrɪðəm/ *n.* 生物周期；生物节律

Tasks

 I. Group Talk

Chinese sayings are a good way to reflect some cultural elements. As for dumplings, we have sayings like "冬至到，吃水饺" "饺子就酒，越喝越有", etc., which reflect the importance of Jiaozi in northern China. Can you think of any other sayings that are related to the food eaten during the twenty-four solar terms? And do you think those sayings make sense? Discuss with your partners and share your opinions.

 II. Writing

The Winter Solstice is one of the most important solar terms, and people from different regions of China will have different customs, especially eating customs, on this day. Find out different eating customs on the Winter Solstice based on your research, and write an article of 80—100 words.

_____

_____

_____

_____

_____

_____

_____

III. Translation

Translate the following sentences into English.

1. 二十四节气是我国上古农耕文明的产物，在农业生产方面起着指导作用。它还影响着古人的衣食住行，甚至是文化观念。

_____

_____

2. 大暑相对于小暑，更加炎热，是一年中日照最多、最炎热的节气。

_____

_____

3. 为防止秋燥，人们可以适当多食用各类蔬菜和水果，要注意不吃或少吃辛辣烧烤食品。

_____

_____

4. 中医认为，人体要适应自然四季变化的规律，保持机体与自然的平衡，才能在一年四季中保持健康。

_____

_____

5. 夏季气候炎热，人的消化功能相对较弱，因此，饮食宜清淡。除此之外，还可以多吃苦菜类蔬菜。

_____

_____

IV. Short Video

Make a short video to introduce one solar term that has not been mentioned in the passage, as well as the eating customs relating to it, and share it with your classmates.

Eating Customs in Chinese Traditional Festivals

Passage B

The seven official public holidays in China now include New Year's Day, the Spring Festival, the Qingming Festival, Labor Day, the Dragon Boat Festival, the Mid-Autumn Festival and National Day, among which the Spring Festival, the Qingming Festival, the Dragon Boat Festival, and the Mid-Autumn Festival are considered as four Chinese traditional festivals. Bringing traditional festivals into national statutory holidays can meet Chinese people's emotional needs, strengthen their **affective commitment** and national identity, and thus promote Chinese cultural **prosperity** and spread Chinese culture worldwide.

Food plays an extremely important role in traditional festivals and Chinese people in different regions have different customs, especially eating customs, in different festivals. Below are mainly eating customs of the Qingming Festival, the Dragon Boat Festival and the Mid-Autumn Festival, the three newly-added festivals of national statutory holidays.

The Qingming Festival, also called Tomb Sweeping Day, is a day traditionally for Chinese people to honor **ancestors**. People offer **sacrifices** and express their **condolence** towards ancestors. Also, people may organize spring outings, fly kites, plant willow trees and taste tea. People in **regions south of the Yangtze River** have the custom of eating sweet green rice balls (or Qingtuan), mainly made from wormwood and glutinous rice flour. People in Shandong Province have the tradition of eating spring onion (whose Chinese pronunciation sounds like "cleverness") and **omelette**, as they believe these two kinds of food can improve eyesight and make people clever and **quick-witted**. People in Xiamen and south Fujian will eat thin pancakes, with all kinds of vegetables (especially celery and Chinese chives) stuffed, to **indicate** a thriving prospect for farm crops, livestock and households.

Falling on the 5th day of the 5th lunar month, the Dragon Boat Festival is a folk custom and culture festival which **integrates worshiping**, celebrating, entertaining and eating. Chinese people have the customs of holding dragon boat races (this is how the festival gets its name), gathering herbs, hanging wormwood and calamus on the door, flying kites, wearing sachets, tying colored silk, drinking realgar wine, and of course, eating Zongzi during this festival, to express their wishes to **counteract** evil forces and **embrace** good luck and blessing. Zongzi in the south is often salty, with pork, ham, sausage, chicken or beef as main fillings, while in the north, Zongzi is often sweet, with red dates, red bean paste or preserved fruit as main fillings.

The Mid-Autumn Festival, which usually falls on the 15th of the 8th lunar month, is a traditional festival for **family reunion**. Since ancient times, people all over China eat mooncakes and appreciate the moon, both of which are round and complete, to express their wishes for completeness and reunion. Mooncake **flavors** differ in different regions of China, as we have Cantonese, Beijing-style, Yunnan-style, and Suzhou-style mooncakes. People in Guangdong Province would worship the moon and offer sacrifices, which include mooncakes, pastries, watermelons, apples, pears, grapes, etc. As **osmanthus** flowers blossom during the Mid-Autumn Festival, people may also eat osmanthus cakes and drink wine fermented with osmanthus flowers, and people in Nanjing have the tradition of eating salted duck, also known as osmanthus duck, as the duck's color and flavor are the best during the time when osmanthus blossoms. People in Jiangsu and Zhejiang tend to eat taro, whose Chinese pronunciation sounds like "good luck to come", and live fresh crabs which just mature at this time.

Eating customs for the above three Chinese traditional festivals may change with time, but Chinese people's enthusiasm for food, the sense of ceremony, and love of customs and cultures behind the festivals will never change.

Vocabulary

custom /ˈkʌstəm/ n. 习惯；惯例；习俗

affective /əˈfektɪv/ adj. 情感的；表达感情的

commitment /kəˈmɪtmənt/ n. 信奉；承诺

prosperity /prɒˈsperəti/ n. 繁荣

ancestor /ˈænsestə(r)/ n. 祖先

sacrifice /ˈsækrɪfaɪs/ n. 牺牲；祭品

condolence /kənˈdəʊləns/ *n.* 哀悼；慰问

regions south of the Yangtze River 长江以南地区

omelette /ˈɒmlət/ *n.* 蛋饼；煎蛋卷

quick-witted /ˌkwɪkˈwɪtɪd/ *adj.* 机敏的

indicate /ˈɪndɪkeɪt/ *v.* 预示；表明

integrate /ˈɪntɪɡreɪt/ *v.* 使完整；成为一体

worship /ˈwɜːʃɪp/ *v.* 崇拜；敬奉

counteract /ˌkaʊntərˈækt/ *v.* 抵消；中和

embrace /ɪmˈbreɪs/ *v.* 拥抱；欣然接受

family reunion 家庭团聚

flavor /ˈfleɪvə/ *n.* 风味；口味

osmanthus /ɒzˈmænθəs/ *n.* 桂花；木犀属植物

Tasks

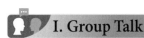

I. Group Talk

Eating mooncakes during the Mid-Autumn Festival and eating Zongzi during the Dragon Boat Festival are common practices for people all over China. Nowadays, flavors, colors and fillings for both mooncakes and Zongzi have changed with improvement of people's living standards and food production technology. Discuss with your partners, summarize the changes of mooncakes and Zongzi, and present your opinions in class.

II. Writing

The passage mainly introduces the eating customs during the three traditional festivals, without touching upon those of the Spring Festival. Research and find out the eating customs for the Spring Festival, paying special attention to differences in northern and southern China. Write an article of 80—100 words.

_____

_____

_____

_____

_____

_____

III. Translation

Translate the following sentences into English.

1. 中国传统节日，是中华民族悠久历史文化的重要组成部分，形式多样、内容丰富。

    _____

    _____

2. 俗话说，清明前后吃艾粄，一年四季不生病。清明节制作清明粄，是客家地区沿袭
    了上千年的习俗。

    _____

    _____

3. 五月五，是端阳；吃粽子，挂香囊；门插艾，香满堂；龙舟下水喜洋洋。

    _____

    _____

4. 中秋节自古便有祭月、赏月、吃月饼、玩花灯、赏桂花、饮桂花酒等民俗。

    _____

    _____

5. 将传统节日定为法定假日，对于弘扬传统文化、促进中华文化的伟大复兴有着深远
    的意义。

    _____

    _____

IV. Short Video

Make a short video about the eating customs of the three traditional Chinese festivals in your hometown, and share it with your classmates and friends.

Passage C

Food for Important Occasions in Life

Throughout life time, Chinese people's important **occasions**, such as birthday party and **wedding ceremony**, all relate to food, as the Chinese saying goes, "Food is the paramount necessity of the people."

Celebration for people's birth and birthday is always important. When a baby is born, and turns to one month old, or 100 days old, or one-year-old, parents will invite friends and relatives to celebrate. Parents will **dress up** the baby, hold home parties, and enjoy big meals with guests. They will also prepare red eggs for guests to bring home. **In return**, guests usually send gifts to the baby to express their best wishes. Moreover, there will be a grabbing test for one-year-old baby called draw lots (or Zhuazhou in Chinese)[1], in which the baby is presented with lots of things in front and the adults will see what the baby grabs to guess at his or her future interest, **personality** and prospect. Since then, each birthday will be celebrated, especially when a person turns to his or her tens, twenties, thirties, etc. In addition to eating birthday cakes on one's birthday, eating long noodles, which conveys **longevity**, has also become a custom for Chinese people.

Getting married is **undoubtedly** a vital event in people's life. A wedding ceremony is usually held in a hotel or in the open air, with guests invited to **witness** and join the ceremony. Before the ceremony, the couple in northern China will eat dumplings, the number of which is usually **even**, as Chinese people believe that good things should be in pairs[2]. The dumpling wrapper is usually made with dragon-fruit juice, as the color red in China conveys happiness. The Chinese name for dumplings, Jiaozi, sound like "Jiaozi" — a **handover** ceremony from parents of the **bride** to the **bridegroom**. In southern China, the couple tend to eat glue puddings (also called sweet dumplings or Tangyuan), which almost mean the same as dumplings in the north. After the wedding ceremony, all guests could **enjoy a square meal**, and drink beer, wine, spirits, soymilk, or soda water as they wish. During the meal, the couple will **make toasts to** guests, who, **in turn**, will express their best wishes to the couple.

For important occasions in people's life, a banquet has become a universal custom practiced by all ethnic groups of China, which almost contains the **essence** of traditional Chinese **dietary culture**. The **banquet** is a dietary form of many people dining together and has been established through long social practice. "Eight Big Bowls"[3], the jargon by folk cooks, are a must in all kinds of banquets. They refer to eight core **main courses** served in big bowls on the table, as "eight" in Chinese has the similar pronunciation with the word "thriving", meaning "to **make a fortune**". Traditionally, the eight main courses contain four meat dishes and four vegetables. With regional characteristics, "Eight Big Bowls" satisfy people with different **social status** and flavors. Nowadays, main courses for "Eight Big Bowls" in both cities and countryside have changed with the improvement of people's living standards, but the custom of people dining together has been kept and will be kept.

All in all, food plays extremely significant roles in Chinese people's life, especially in important occasions. The **folk wisdom** tells you that nothing cannot be handled by a meal; if not, have another one.

Notes

[1] draw lots (or Zhuazhou in Chinese) 抓周，小孩周岁时举行的一种预测前途和性情的仪式。

[2] Good things should be in pairs. 好事成双。

[3] Eight Big Bowls 八大碗（用于宴客之际，每桌八个人，桌上八道菜，上菜时都用统一的大海碗，摆放成八角形）

Vocabulary

occasion /əˈkeɪʒn/ *n.* 重大场合
wedding ceremony 婚礼
celebration /ˌselɪˈbreɪʃn/ *n.* 庆祝

dress up 盛装打扮
in return 作为回报
personality /ˌpɜːsəˈnæləti/ *n.* 性格；个性

longevity /lɒnˈdʒevəti/ *n.* 长寿；寿命

undoubtedly /ʌnˈdaʊtɪdli/ *adv.* 毫无疑问地

witness /ˈwɪtnəs/ *v.* 目击；见证　*n.* 证人

even /ˈiːvn/ *adj.* 偶数的；均衡的

handover /ˈhændəʊvə(r)/ *n.* 交接；移交

bride /braɪd/ *n.* 新娘

bridegroom /ˈbraɪdgruːm/ *n.* 新郎

enjoy a square meal 饱餐一顿

make toasts to 祝酒；敬酒

in turn 相应地；转而

essence /ˈesns/ *n.* 本质；实质

dietary culture 饮食文化

banquet /ˈbæŋkwɪt/ *n.* 宴会 *v.* 宴请

main course 主菜

make a fortune 发财；赚大钱

social status 社会地位

folk wisdom 民间智慧

Tasks

 ## I. Group Talk

It is said that the diet for a baby's mother, especially during the period of confinement, is quite exquisite in all places of China. Research and discuss with your partners on the lying-in woman's diet in different places of China. What do they need to pay special attention to? And how has their diet in the past changed as time passes by?

 ## II. Writing

Nowadays, the graduation feast has aroused heated discussion in the society. Some believe it is a good way to announce good news of students' being admitted into university and to express students' gratitude towards teachers, relatives and friends for their help during the school years. But some hold that the graduation feast has become a method to accumulate wealth and may involve bribe, resulting in materialization and vulgarization, and thus should be abolished. Write an article of 80—100 words to illustrate your opinion with solid evidence.

_____

_____

_____

_____

_____

III. Translation

Translate the following sentences into English.

1. 饮食活动深刻地影响社会上的礼仪、风俗、庆典、战争、婚丧嫁娶等，与各式各样的社会活动有着密切紧凑的互动联系。

   _____

   _____

2. 中国的酒文化历史源远流长，很多典籍中都有关于酒和饮酒文化的记载，酒文化深入中国人的血脉深处，影响深远。

   _____

   _____

3. 满月酒，是指婴儿出生后一个月而设立的酒宴，目的是庆祝婴儿渡过了第一个月的难关。

   _____

   _____

4. 生日吃的面条被称为"长寿面"。一般来说，长寿面整碗只有一根面条，吃的时候最好不要弄断。

   _____

   _____

5. 八大碗在当时集中了扒、焖、酱、烧、炖、炒、蒸、熘等所有的烹饪手法。

   _____

   _____

IV. Short Video

Make a short video to introduce or present an important occasion of feast that you witness and participate, and share it with your classmates.

Supplementary Reading

Food at Christmas

Christmas Day, on December 25, is a public holiday in many, but not all, countries, such as the US, Canada, Australia, and many European countries. People celebrate Christmas Day in various ways. Festive activities include decorating Christmas trees, holding Christmas parties, exchanging Christmas gifts, and singing Christmas songs, etc. It is also a special time for children to receive gifts from their parents, relatives, and the mythical figure Santa Claus. On Christmas Eve, the evening just before Christmas, Santa Claus is said to enter children's home and fill their stockings hanging at the end of their bed or shoes put beside the fireplace, with gifts and sweets.

In most countries, it is a common practice to organize a special meal on Christmas Day, often consisting of turkey and a lot of other festive foods, for family or friends. Turkey is a traditional main course and the most distinguished dish eaten in America and many other countries. Roast turkey is not difficult to make. Usually, the turkey is filled with all kinds of seasonings and well-mixed stuffing in the belly, and roasted as a whole until its skin turns deep brown. Then it is cut into thin slices, and topped with gravy and salt to one's taste. With shiny skin, and tender and smooth meat, the cooked turkey both smells good and tastes good.

Traditional English roast uses turkey, goose or chicken as main ingredient, supplemented by vegetables like bean sprouts, green pepper, carrot, celery, potato, and onion, and topped with tasty gravy. Other Christmas foods British people enjoy include roast pork, dry fruit pie, and Christmas pudding. Well-received in most

European countries, Christmas pudding is made from cream, vanilla, and cinnamon, with almond hidden in it. It is usually a routine for the family to find almond in the pudding after enjoying Christmas feast, and the lucky dog, who finds or eats the almond, would get a prize prepared by the family member.

France, well-known for its fine food, has goose liver, oyster, truffle, chestnut, and all kinds of cakes and pastries for Christmas, among which Buche de Noel[1] is a must. It is a cake made in the shape of a firewood or a tree trunk. One legend goes that burning oak in the fireplace protects people from diseases and disasters. Another tale tells that a poor young man, who can't afford to buy a Christmas gift for his girlfriend, sends a piece of firewood instead, and wins her heart. Later on, people make Buche de Noel to symbolize good luck.

Ginger biscuit (or Lebkuchen[2]) is the most famous Christmas food in Germany, yet has spread to some other European countries. Traditionally made of honey and pepper corn, ginger biscuit is hot and spicy in taste, with a layer of icing on the surface to make it multi-flavored and beautifully-shaped. Christmas goose, often roasted, is also a traditional food for German people.

Christmas in Australia falls in summer, so more Australian people tend to barbecue outdoors, instead of in the fireplace. Ham and seafood are very popular ingredients for barbecue. The most popular Christmas dessert in Australia is called White Christmas, which mixes ingredients such as raisin, preserved cherry, dry coconut, icing, milk powder, and coconut oil.

People in many other countries, not listed above, also celebrate Christmas in their own way. No matter how people celebrate this festival, no matter what food they eat, Christmas is a great time to enjoy together with family and friends.

Notes
[1] Buche de Noel 法国树干蛋糕
[2] Lebkuchen 德国姜饼

Recipe Sharing

Qingtuan with Red Bean Paste

Ingredients

100g wormwood, 60g wheat starch, 2g baking soda, 200g glutinous rice flour, 100g water (used to boil wormwood), 60g water (used to scald wheat starch), 350g red bean paste, 10g corn oil

Preparation

1. Pick wormwood, clean thoroughly and let dry. Add 100g wormwood and 2g baking soda into 100g water, boil till soft, then put into a food processor, and make a wormwood paste.

2. Boil 60g water, quickly pour into 60g wheat starch, scald to make it transparent, then mix with 200g glutinous rice flour and wormwood paste to make a dough. Knead the dough after adding 10g corn oil.

3. Divide red bean paste into portions of 25g, roll each portion round. Take a 40g dough to wrap up the fillings, and roll round, too.

4. Boil water and steam Qingtuan in a steamer till cooked.

豆沙青团

食材配方

艾草100克、澄面60克、小苏打2克、糯米粉200克、清水100克（煮艾草用）、清水60克（烫澄面用）、豆沙馅350克、玉米油10克

制作工艺

1. 艾草采摘，洗净晾干。100克清水加入100克艾叶和2克小苏打煮软，放入料理机打成艾草糊。

2. 60克清水烧开，迅速倒入60克澄面中，将澄面烫熟，成透明状，与200克糯米粉和艾草糊一起混合成面团。将10克玉米油揉进面团。

3. 将红豆馅分割成25克一个搓圆，待用。取40克面团将豆沙馅包裹住，搓圆。

4. 水烧开，放入青团隔水蒸15分钟，美美的青团就可以出炉了。

Recipe Sharing

Christmas Pudding

Ingredients

30g dried cranberries, 25g dried blueberries, 25g dried blackcurrants, 100g rum, 50g almonds, 50g walnuts, 25g almond powder, 25g low gluten flour, 2 eggs, 1 piece of toast, 50g brown sugar, 1 lemon, cherries in proper amount, light cream in proper amount

Preparation

1. Add dried cranberries, dried blueberries, dried blackcurrants into a bowl and pour rum into the bowl to soak for one evening.

2. Pour the dried fruits in a big bowl, add 2 eggs, almond, walnuts, almond powder and low gluten flour. Shred the toast, put it into the bowl, and then add brown sugar. Finally add lemon peel into the bowl.

3. Stir well the ingredients in the bowl, and get pudding paste. Put it into another bowl, press to make it compact, steam over high flame in a steamer for 45 minutes after the water boils. Top with light cream and decorate with cherries.

圣诞布丁

食材配方

蔓越莓干30克、蓝莓干25克、黑加仑干25克、朗姆酒100克、扁桃仁50克、核桃仁50克、杏仁粉25克、低筋面粉25克、鸡蛋2个、吐司1片、红糖50克、柠檬1个、樱桃适量、淡奶油适量

制作工艺

1.碗中倒入蔓越莓干、蓝莓干、黑加仑干、朗姆酒浸泡一夜。

2.将浸泡过一夜的果干倒入一个大碗中,接着倒入2个鸡蛋、扁桃仁、核桃仁、杏仁粉和低筋面粉。将吐司撕碎放入碗中,接着倒入红糖。将柠檬削皮,取柠檬皮放入碗中。

3.将碗中的食材翻拌均匀,做成布丁糊,再将翻拌均匀的布丁糊放入碗中压实,放入烧开水的蒸锅大火蒸45分钟。淋上淡奶油,放上樱桃做装饰就完成了。

CHAPTER III DINING ETIQUETTE

Chinese Dining Etiquette and Table Manners

Passage Ⓐ

China's dining etiquette, as an essential part of the sophisticated "Li" system, can be traced back to Zhou Dynasty (1046—256 BCE). Priding themselves on being **descendants** of the "nation of etiquette", Chinese people attach great importance to sets of rules that govern their behaviors at the table. Though the regulations change with time and vary from region to region, three notions behind them remain eternal—courtesy, **hospitality** and elegance.

A polite social interaction begins with the reception. When the guests show up, warmest greetings from the host and hostess will be presented and the guests will be led all along to their seats. Seating arrangement is a manifestation of **hierarchical** or **patriarchal** structures deeply-rooted in Chinese society. The seat facing the entrance of the room or the one in the center facing east of the room is regarded as the "seat of honor", which is taken by the "guest of honor" — the most highly ranked person or the most senior one among those in attendance.

When everyone is seated, the host and hostess will propose the first toast to appreciate the presence of all the guests and to give special thanks to the guest of honor. After the first toast, when the guest of honor says words to the effect of "Let's eat", and raises his chopsticks, the guests will know it is time to **commence** dining.

As the old saying goes, "No alcohol, no **protocol**." Alcoholic drinks, light or strong, are often served throughout the meal. To communicate generosity, the host will always make sure everyone's glasses are not empty for long.

Toasts to each other at the banquet table will be made in order of **seniority**. If a guest wishes to give a toast, he should first drink to the superiors or the elders. It is also worth noting that the younger member should hold his glass with both hands and **clink** the **rim** of his glass below that of an elder's to show politeness and respect. Toasts to the host and hostess are also necessary to show one's gratitude for being invited. The **utterance** of **auspicious** words, together with the clinking of glasses, makes the feast proceed in a joyous and harmonious atmosphere.

When dining at a Chinese banquet, **proprieties** are never supposed to be neglected. The dishes, usually in ample supply, are placed on a **lazy Susan** — a round revolving disc in the center of a big round table — so that every diner has equal access to the dishes, no matter where he sits. It is considered disrespectful to rotate the lazy Susan while someone else on the table is serving himself from the main bowl. Besides, if someone wants to have a certain dish for the second time, it is recommended to wait until the dish takes a complete round that everyone gets his due share.

In the past, it was acceptable for hospitable hosts or caring elders to get food with their own chopsticks for the guests or kids. But the same conduct is deemed inappropriate today. Out of consideration for health and **hygiene**, now each diner would have two pairs of differently colored chopsticks. One is served as "Gongkuai", or sharing chopsticks, to pluck food from **communal** dishes to put into individual bowls, and the other is for eating from individual bowls. It is okay to offer guests with food using a pair of sharing chopsticks. However, allowing the guests to help themselves would be better.

The Chinese dining table is where relationships are built and maintained. Good dining etiquette and table manners can show one's good self-cultivation and leave a good impression on others. For Chinese people, you are how you eat.

Vocabulary

etiquette /ˈetɪket/ *n.* 礼仪

descendant /dɪˈsendənt/ *n.* 后代

hospitality /ˌhɒspɪˈtæləti/ *n.* 好客

hierarchical /ˌhaɪəˈrɑːkɪkl/ *adj.* 等级制度的

patriarchal /ˌpeɪtriˈɑːkl/ *adj.* 家长制的

commence /kəˈmens/ *v.* 开始

protocol /ˈprəʊtəkɒl/ *n.* 礼节

seniority /ˌsiːniˈɒrəti/ *n.* 年长

clink /klɪŋk/ *v.* 碰杯

rim /rɪm/ *n.* 边缘

utterance /ˈʌtərəns/ *n.* 说出；表达

auspicious /ɔːˈspɪʃəs/ *adj.* 吉祥的

proprieties /prəˈpraɪətiz/ *n.* 礼节；行为规范

lazy Susan （餐桌上盛食物便于取食的）
圆转盘

hygiene /ˈhaɪdʒiːn/ *n.* 卫生

communal /kəˈmjuːnl/ *adj.* 公用的

Tasks

 ## I. Group Talk

What do you think of the following quotations? Discuss them in groups.

1. The way you cut your meat reflects the way you live.

— Confucius

2. Manners are a sensitive awareness of the feelings of others. If you have that awareness, you have good manners, no matter what fork you use.

— Emily Post

 ## II. Writing

The Chinese dining table has witnessed many changes in concepts, attitudes and behaviors concerning dining etiquette. For example, it is not uncommon to see people, usually the middle-aged, to fight over the bill in a restaurant. For them, offering to pay the bill is a way to show their generosity, politeness and sincerity. However, the young generation is more likely to split the bill, and arguing over the bill in public makes them feel embarrassed. What do you think about this change? Do you know some other changes to traditional dining etiquette? Write an article of about 80 words to express your opinions.

_____

_____

_____

_____

_____

III. Translation

Translate the following sentences into English.

1. 食不言，寝不语。

2. 在中国宴会上，座次讲究"尚左尊东""面朝大门为尊"。

3. 据文献记载，至少在周代，饮食礼仪已形成一套相当完善的制度，是历朝历代表现大国之貌、礼仪之邦、文明之所的重要方面。

4. 有的人吃饭喜欢使劲咀嚼又干又脆的食物，发出很清晰的声音来，这种做法是不合礼仪要求的。

5. 离席时，宾客应向主人表示感谢，也可在此时邀请主人以后到自己家做客，以示回敬。

IV. Short Video

Make a short video about traditional Chinese dining etiquette and table manners not mentioned in the passage, and share it with your classmates.

Folk Beliefs about Dining

Passage Ⓑ

In China, there are **superstitious convictions** and practices **galore** connected with daily activities. Dining-related **folklore**—on the subject of food, utensils, table manners and eating taboos—has been passed down through generations by oral tradition to teach people what to do and what to prevent.

Chinese people lay great emphasis on "**supernatural**" **properties** of numbers. In southern China, sweet glutinous rice dumplings, called Tangyuan, are served both on the first day of Chinese New Year and on the Lantern Festival. The number of Tangyuan in a bowl, usually even, carries different meanings. For example, the figure two means good things come in double; six signifies that everything goes smoothly; eight stands for extra prosperity. There is a controversy regarding the connotation carried by number four. **Partaking of** four Tangyuan can be interpreted as staying healthy in all the four seasons, or making a good fortune all the year round. However, since four has the similar pronunciation with the word "death" in mandarin, most people will consciously avoid it.

The Chinese are also cautious with their food containers and eating tools, especially rice bowls and chopsticks, the two indispensable utensils used for daily meals. In Chinese language, a rice bowl has the **metaphorical** meaning of one's employment. Thus, **shattering** a bowl **portends** the loss of one's livelihood or the arrival of unexpected **calamities**. In order to turn ill luck into good, people will say the blessing "Sui Sui Ping'an", indicating "safe and sound every year", since the original "Sui Sui", meaning "break into pieces", is **homophonous** with the word meaning "year by year". Some people also believe that if they wrap the broken pieces in a red cloth before disposing of them, the effect of the **frightful presage** can be **counterbalanced**. Inappropriate use of chopsticks will be regarded as being ungraceful, offensive and a lack of family education. Sticking chopsticks **vertically** into a bowl of rice is definitely prohibited, since chopsticks standing up right in the rice resembles burning **incense** placed in the **censer** in a mourning hall, which reminds people of death. **Banging** the edge of a bowl or a plate with chopsticks will be frowned upon as this act is **reminiscent** of beggars attracting attention from passersby to ask for meals or money, and begging is seen as **undignified** and **humiliating** by the Chinese.

Aside from implied meaning of numbers and proper use of tableware, special attention is paid to the way people eat certain food. Sharing is reckoned a traditional virtue in China, but do not attempt to share a pear with friends or relatives. The act of dividing a pear is pronounced as "Fen Li" in Chinese, which is a homophone to another word meaning "departure" (or to make it a **pun**, "dividing a pair"). For this reason, it is assumed that sharing a pear will cause friendship splitting, families separating, and even couples getting a divorce. Fish, Yu, sounding like the word for abundance, is usually served whole in a feast. After finishing one side of the fish, rather than flipping it to the other side, it is better to carefully remove the backbone and then get to the flesh underneath. People living in coastal areas consider turning over a fish is equivalent to turning over a fishing boat, which is a huge misfortune to a fisherman's family. In Sichuan Province, children are raised on the saying or, to be accurate, the warning told by their old grandmas that they must finish every grain of rice in their bowls or they will marry a spouse with **pockmarks** when they grow up.

The above dining beliefs might be considered as irrational and ridiculous in present days, but there are still good reasons to preserve them, for they can reflect people's expectations for a better life, awe for nature and inheritance of traditional morals.

Vocabulary

superstitious /sjuːpəˈstɪʃəs/ *adj.* 迷信的

conviction /kənˈvɪkʃn/ *n.* 确信；信仰

galore /gəˈlɔː(r)/ *adj.* 大量的

folklore /ˈfəʊklɔː(r)/ *n.* 民间风俗

supernatural /ˌsjuːpəˈnætʃrəl/ *adj.* 超自然的

property /ˈprɒpəti/ *n.* 属性

partake of 吃；喝

metaphorical /ˌmetəˈfɒrɪkl/ *adj.* 比喻的

shatter /ˈʃætə(r)/ *v.* 打碎

portend /pɔːˈtend/ *v.* 预示

calamity /kəˈlæməti/ *n.* 灾祸

homophonous /həˈmɒfənəs/ *adj.* （词语）

同音异形的；同音异义的

frightful /ˈfraɪtfl/ *adj.* 可怕的

presage /ˈpresɪdʒ/ *n.* 预兆

counterbalance /ˌkaʊntəˈbæləns/ *v.* 抵消

vertically /ˈvɜːtɪkli/ *adv.* 垂直地

incense /ˈɪnsens/ *n.* 香

censer /ˈsensə(r)/ *n.* 香炉

bang /bæŋ/ *v.* 敲击

reminiscent /ˌremɪˈnɪsnt/ *adj.* 引人联想的

undignified /ʌnˈdɪɡnɪfaɪd/ *adj.* 无尊严的

humiliating /hjuːˈmɪlieɪtɪŋ/ *adj.* 丢脸的

pun /pʌn/ *n.* 双关语

pockmark /ˈpɒkmɑːk/ *n.* 麻子

irrational /ɪˈræʃənl/ *adj.* 不合理的

inheritance /ɪnˈherɪtəns/ *n.* 继承

Tasks

 I. Group Talk

Chinese people assign special meanings to ordinary foods based on their characteristics such as shape, flavor, color, name, etc. For example, it is a tradition to have noodles on one's birthday since long noodles stand for longevity; pomegranates, the fruit filled with vibrant red seeds, are viewed as a symbol of fertility, which is important in traditional Chinese culture. Can you think of more examples? Talk with your partners and share your opinions.

 II. Writing

Chinese people have been using chopsticks for more than 3,000 years and have developed sets of rules associated with the use of chopsticks. Under the cultivation of Confucianism, which advocates gentleness and benevolence, Chinese people have banned violent actions from the dining table, and hence it is rude to pick up food by stabbing it with chopsticks. Collect more information about proverbs, rules or traditions about chopstick etiquette and write an article of about 80 words.

_____

_____

_____

_____

_____

III. Translation

Translate the following sentences into English.

1. 中餐礼仪要求大家做到坐有坐相，吃有吃相。

_____

_____

2. 中国人认为饮食与个人命运息息相关，用餐时若犯了禁忌，则可能惹来厄运。

_____

_____

3. 通常来说，在新年期间吃白色的东西是不吉利的，白色寓意不祥甚至是死亡。但是在一些中国人的眼里，白色也意味着纯洁。

_____

_____

4. 中国新年最普遍的水果是橘子。"橘"与"吉"谐音，且橘子外皮是耀眼鲜明的橘色，这是一种象征着财富的颜色。

_____

_____

5. 在腊月二十三，也就是小年这一天，人们会买糖瓜供奉灶王爷，祈求灶王爷嘴甜些，上天言好事。

_____

_____

IV. Short Video

Make a short video about folk dining beliefs in your hometown, and share it with your classmates.

A Sip of Sichuan

Passage C

If a Chinese is asked to describe what life is, odds are that he will answer with the old saying, "Life is what consists of firewood, rice, oil, salt, soy sauce, vinegar and tea." Tea is regarded as one of the seven necessities people cannot live without. Having been consumed as the most popular beverage for thousands of years, tea can be found in different occasions for different purposes, including daily intakes, festival celebrations, business receptions, sacrificial offerings and so forth.

China is widely known as the hometown of tea, and Sichuan is said to be the first place in China to grow and process tea. Sichuan Basin, located in the southwest of China, is endowed with fertile farmlands and humid subtropical climate, which makes it suitable for tea cultivation. People of Sichuan consume a particularly **copious** amount of tea every year and Sichuan Province possesses a large number of teahouses far in excess of any other areas.

In Sichuan, the most common teaware for **steeping** and drinking tea is called "Gaiwan", a three-piece utensil made of **ceramic** or **porcelain**. A Gaiwan is also called "San Cai Wan" (directly translated as "three-talent cup") as the three components — the lid, the cup and the **saucer** — together embody the central idea of Taoism, that is, the harmony among heaven, man and earth. In teahouses, the Sichuan kung fu tea ceremony — the whole **gamut** of kettle tricks combining **martial arts** and **acrobatics** — is a popular performance among both locals and tourists. By pouring boiling water into customers' Gaiwan from a 1.2-meter-long spout copper kettle without **splashing** a drop, the tea master provides a feast both for the mouth and the eyes.

A teahouse is the **microcosm** of a society. Men and women, old and young, from every stratum of society, with distinct life stories, all come to the teahouse to order a cup of tea that fits their own budgets. Tea drinking is a mass behavior, but rules concerning tea drinking are rather **niche** and complicated, so much so that only **veteran** tea drinkers can perform and understand. To them, a Gaiwan can talk. If noises are made while removing the lid from the cup, surely the Gaiwan user is expressing his anger or **discontent**. If the lid is placed leaning on the saucer, a refill is needed. If a tree leaf is put on the lid and no one is in the seat, it is a sign reminding the waiter not to clear away the tea sets and the customer will come back soon.

Unfamiliarity with the implicit messages communicated in a teahouse is absolutely understandable, but on important days, correct **observance** of etiquette concerning tea serving and drinking must be ensured. On a traditional Sichuan wedding day, a tea ceremony is conducted and the bride and groom will serve tea to their parents and parents-in-law. The new couple, dressed in traditional wedding garments, will kneel down on a pair of cushions and each one will hold a Gaiwan of tea high above his or her head. This is the way to show their respect for and **filial piety** to the older generation. Gaiwan used for a wedding is painted red, with the Chinese character meaning "double happiness" printed on it. After accepting the Gaiwan and taking a sip of tea, the parents will give good wishes to the newly-weds, as well as a red packet. Acceptance of the tea means the consent of the marriage.

As an essential **constituent** part of Sichuan culture, tea-drinking **epitomizes**, as nothing else can, the soul of this province. The wisdom of Sichuan people, the mysterious oriental philosophy and the unique characters of cities and towns in this province can all be tasted in a Gaiwan of tea.

Vocabulary

sip /sɪp/ *n.* （抿）一小口

copious /ˈkəʊpiəs/ *adj.* 大量的

steep /stiːp/ *v.* 浸；泡

ceramic /səˈræmɪk/ *n.* 陶

porcelain /ˈpɔːsəlɪn/ *n.* 瓷

saucer /ˈsɔːsə(r)/ *n.* 茶托

gamut /ˈɡæmət/ *n.* 全范围；全过程

martial arts *n.* 武术

acrobatics /ˌækrəˈbætɪks/ *n.* 杂技

splash /splæʃ/ *v.* 溅起水花

microcosm /ˈmaɪkrəʊkɒzəm/ *n.* 缩影

niche /niːʃ/ *adj.* 小众的

veteran /ˈvetərən/ *adj.* 经验丰富的

discontent /ˌdɪskənˈtent/ *n.* 不满

observance /əbˈzɜːvəns/ *n.* 遵守

filial piety *n.* 孝顺

constituent /kənˈstɪtʃuənt/ *adj.* 构成的

epitomize /ɪˈpɪtəmaɪz/ *v.* 体现

Tasks

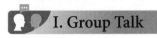 **I. Group Talk**

What do you think of the following quotations? Discuss them in groups.

1. There is something in the nature of tea that leads us into a world of quiet contemplation of life.

— Lin Yutang

2. Tea tempers the spirits and harmonizes the mind, dispels lassitude and relieves fatigue, awakens thought and prevents drowsiness, lightens or refreshes the body, and clears the perceptive faculties.

— Lu Yu

 II. Writing

Chinese tea ceremony, or "Chadao", is not simply drinking of tea. It is the combination of brewing, smelling, drinking and appreciation of tea. Learn more about Chinese tea ceremony steps and etiquette and write an article of about 80 words.

_____

_____

_____

_____

_____

_____

_____

_____

_____

III. Translation

Translate the following sentences into English.

1. "客来敬茶"，这是中国汉族最早重情好客的传统美德与礼节。宾客到家，总要沏上一杯香茗。

   _____

   _____

2. 民间男女订婚以茶为礼，女方接受男方聘礼，叫"下茶"，并有"一家不吃两家茶"的谚语。

   _____

   _____

3. 酒满敬人，茶满欺人。

   _____

   _____

4. 旧社会的茶馆还兼有调解社会纠纷的职能。亲朋邻里之间若出现了纠纷，双方约定到某茶馆"评理"。

   _____

   _____

5. 从四川茶馆或喧闹嘈杂、或平和闲适之中，也许能体味到一些巴蜀民风、捕获一些市井之趣。

   _____

   _____

IV. Short Video

Visit a Sichuan teahouse and make a video to introduce it. Share the video with your classmates.

Supplementary Reading

British Afternoon Tea Etiquette

> Under certain circumstances there are few hours in life more agreeable than the hour dedicated to the ceremony known as afternoon tea.
>
> — Henry James

Brits are known for their love of tea. When tea arrived in British shores from China in the mid-17th century, this aromatic beverage quickly gained its popularity among aristocrats and then became available to everyone. About two hundred years later, the British developed the afternoon tea tradition and made it the most quintessential of British customs.

It is assumed that, Anna Maria Russell, the 7th Duchess of Bedford, first practiced the concept of afternoon tea. Back to those days, it was customary for the high-society to have only two meals a day—breakfast in the morning and dinner served around 8 p.m. — leaving a long period of time between them. Getting peckish in the late afternoon, the Duchess asked a light meal of snacks accompanied by a cup of tea to be brought to her to assuage hunger. She then began to invite friends to join her at her little tea parties and later on such private home gatherings became fashionable among society women. It was when Queen Victoria participated in those events that afternoon tea has evolved into a formal social occasion and a national pastime.

The other common name for afternoon tea is "low tea", since it was most often taken by the aristocracy in a relaxed manner at a low table, like a coffee table or a tea table in the living room. Caution! Do not call it "high tea", or one will be taken as ignorant and ill-bred. High tea somehow sounds grander than afternoon tea, but it actually refers to the early evening meal taken by servants at higher dinner tables after laborious work, not because it involves high class.

Etiquette dictates that one person is assigned to pour the tea for everyone at the table. Once tea has been poured in the fine china cups for all the guests, dressings like milk, sugar or lemon can then be added. When stirring his tea, one should gently sweep the spoon back and forth two or three times, instead of stirring in a circular motion. When drinking the tea, one should sit straight, hold the teacup by its handle properly (avoid extending the little finger to balance the cup), lift the cup to lips, and take small sips.

The traditional English afternoon tea consists of an elegant selection of snacks presented on a three-tiered cake stand, with finger sandwiches on the first layer, scones on the second, cakes or pastries on the third. There is an order in which the accompaniments should be eaten during the tea time: from top to bottom. When eating a scone, remember to break it in half by hand, then spread jam and cream onto it and eat each half separately. It is considered rude to cut the scone with a knife, or just take bites out of the whole thing.

There are few things in the world being more British than the tradition of afternoon tea. Although the young British nowadays are not that strict about how others drink their tea, afternoon tea is still an important business in Britain to forge friendships. Proper etiquette is definitely a sign of respect to both the host and other guests, and can help one create a good image.

Recipe Sharing

Fried Rice Balls

Ingredients

150g glutinous rice flour; 25g wheat starch; 60g white sugar; 60g brown sugar; 20g white sesames; 60g water; 2000g rapeseed oil

Preparation

1. Add hot water at 95℃ to wheat starch and stir thoroughly. Add glutinous rice flour, water and white sugar to wheat starch mixture. Knead into a dough.

2. Shape the dough into a long bar and cut into portions. Each portion weighs approximately 60g.

3. Shape each portion into a ball. Sprinkle dough balls with white sesame seeds.

4. Heat a wok over flame. Add rapeseed oil and heat to 150℃. Deep fry the dough balls until they float on the surface.

5. Add brown sugar and stir constantly with a ladle. Keep frying until the dough balls increase in volume and turn golden brown. Transfer to a serving dish.

糖油果子

食材配方

糯米粉 150 克、澄面 25 克、白糖 60 克、红糖 60 克、白芝麻 20 克、清水 60 克、菜籽油 2000 克

制作工艺：

1. 澄粉用 95℃水温烫揉均匀，加入糯米粉、清水、白糖揉匀成团。

2. 将面团搓成长条后切分成剂子，每个剂子约 60 克。

3. 将剂子搓圆，裹上白芝麻，制成生坯。

4. 锅置火上，入油烧至 150℃，下生坯炸至浮面。

5. 下红糖，用炒勺不断搓动，炸至膨胀、呈金黄色捞起，装盘。

Recipe Sharing

Cucumber Sandwiches

Ingredients

1 cucumber; 1 package of cream cheese (8 ounce), softened; 1/4 cup of mayonnaise; 1/4 teaspoon of garlic powder; 1/4 teaspoon of onion salt; a dash of Worcestershire sauce; a loaf of sliced bread (1 pound), crusts removed; a pinch of lemon pepper

Preparation

1. Peel and thinly slice the cucumber; put cucumber slices between 2 paper towels for about 10 minutes to drain the liquid.

2. Mix cream cheese, mayonnaise, garlic powder, onion salt, and Worcestershire sauce in a bowl until smooth.

3. Spread cream cheese mixture evenly on one side of each bread slice.

4. Lay cucumber slices over half of the bread slices; sprinkle with lemon pepper.

5. Top with a second slice of bread; cut into triangles.

黄瓜三明治

食材配方

黄瓜 1 根、奶油奶酪 1 袋（8 盎司，已软化）、蛋黄酱 1/4 杯、蒜粉 1/4 茶匙、洋葱盐 1/4 茶匙、伍斯特沙司少许、去皮切片面包 1 条（1 磅）、柠檬胡椒一撮

制作工艺：

1. 黄瓜去皮切薄片，放置在厨房纸上 10 分钟，排出多余水分。

2. 将已软化的奶油奶酪、蛋黄酱、大蒜粉、洋葱盐、伍斯特沙司混合均匀。

3. 将奶酪糊均匀抹在每片面包上。

4. 取一半的面包，铺上黄瓜片，撒柠檬胡椒。

5. 取剩下的面包盖在黄瓜上，做成三明治；沿对角线切开，呈三角形。

KITCHENWARE

Chinese Kitchen Utensils

Passage Ⓐ

There are so many secrets in the Chinese kitchen, with cookware being one. The Chinese people create different utensils, and choose to use them differently while handling various ingredients. A wide variety of cooking techniques have been developed based on these diversified utensils, but the later could facilitate the creation of a better flavor. While new **cookware** could be found in most Chinese kitchens, many traditional utensils are still widely used today.

The **pottery** cookware is not the star player in the modern kitchen, but there are two pieces of pottery utensils, namely pottery **casseroles** and **jars**, which are still working, like living specimens. The casserole is the favorite pot for making **stews** in China, and the most amazing creation of Chinese utensils is the pickle jar, which can be found almost everywhere in Sichuan. Most of the jars are made of **clay**. The thin body of the jar makes it easy for the **microorganism** to breath — even after a long time of pickling, the vegetable could still be fresh, with **crisp** and **refreshing** taste. It is a wonder who invented such a simple and scientific pickle jar: pour a bowl of water on the mouth of the jar, **buckle** a large bowl on the top, and the pickle in the jar will be in the **anaerobic fermentation**.

A great many of bamboo and wooden utensils are still used in the modern Chinese kitchen, such as bamboo chopsticks, shawls, **steamers**, baskets and wooden cutting boards, molds, bowls, wok scoops, chopsticks, etc. The Chinese were the first in the world to steam food, most commonly using the steamer. The steamed food maintains its full shape, flavor and its nutrition. From a dish to a pastry, the steaming technique retains the original taste for the Chinese. Besides, although the cutting board is of a low profile in the Chinese kitchen, those made of ginkgo wood are a beloved treasure in many homes. The ginkgo cutting board is antibacterial, and its tough surface resists stains and knife marks. When cutting ingredients, the blade would not be wrecked or get slippery on the board.

In addition to a large number of traditional utensils, there are also many modern kitchen appliances in Chinese kitchen now, such as induction cookers, microwave ovens, electric rice cookers, kitchen ventilators, dishwashers, etc. From traditional to modern Chinese kitchens, they are full of our familiar utensils for cooking, serving, eating, drinking and storing. They bear the trace of daily life and history, and accompany each of us to grow up while serving us.

Vocabulary

utensil /juːˈtensl/ n. 器皿

cookware /ˈkʊkweə(r)/ n. 炊具；厨房用具

pottery /ˈpɒtəri/ n. 陶器

casserole /ˈkæsərəʊl/ n. 砂锅

jar /dʒɑː(r)/ n. 坛子

stew /stjuː/ n. 炖菜

clay /kleɪ/ n. 黏土；陶土

microorganism /ˌmaɪkrəʊˈɔːgənɪzəm/ n. 微生物

crisp /krɪsp/ adj. 脆的

refreshing /rɪˈfreʃɪŋ/ adj. 清爽的

buckle /ˈbʌkl/ v. 扣住

anaerobic fermentation 厌气发酵；无氧发酵

steamer /ˈstiːmə(r)/ n. 蒸笼

Tasks

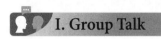 I. Group Talk

Share special kitchen utensils of your hometown and introduce their uses to your classmates.

 II. Writing

Write an article of about 150 words to introduce differences between modern and traditional Chinese kitchens.

_____

_____

_____

_____

_____

_____

_____

_____

_____

_____

_____

_____

_____

III. Translation

Translate the following sentences into English.

1. 2010 年，联合国教科文组织授予四川省省会成都亚洲第一个"美食之都"的称号，川菜终于开始得到全球的瞩目，而它也实至名归。

_____

_____

2. 成都曾是古蜀国的首都，从烹饪的角度讲，至今也仍然算得上"川菜之都"。

_____

_____

3. 传统川菜厨房中使用的设备简单到令人震惊。一把菜刀、一块木菜墩、几个装食材的盘子和碗，备酒菜只需要这些而已。

_____

_____

4. 传统的菜墩是用一截厚实的圆形树干制作的，加盐和食用油来养护，可以一直用很多年。

_____

_____

5. 筷子是日常生活中经常用到的一种进餐工具，也是中国人进餐时的必备用具。

_____

_____

IV. Short Video

Search for utensil-related idiomatic stories. Make a short video to explain one of them, and share with your classmates.

Chinese Cleaver

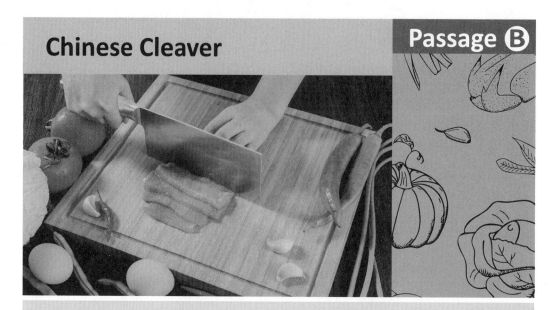

In the Chinese kitchen, the essential knife is the Chinese **cleaver** (Caidao). It is a versatile tool: its **blade** can slice and chop various meat and vegetables, peel ginger, or even **debone** a duck; the corner of its back can crack open a fish-head to release its flavors; its **flat** can crush spring onions, ginger and garlic with a **whack**; turning the knife upside down, the blunt back edge of the blade can **mince** raw meat or fish into a **paste**; holding the cleaver flat, you can **shovel up** the chopped vegetables and put them into the wok. In one word, the Chinese cleaver can chop, **slash**, slice and **dice**, and it is specially formed for the Chinese cuisine.

Cleavers are made of **carbon steel** or **stainless steel**. The secret of hand-making a Chinese cleaver lies in **smashing** against the steel, **quenching** and sharpening the blade. When the steel is smashed into the iron, an indestructible blade will be made. The salty water from the deep well is cold and calm, which is used to quench the cleaver to make it hard on the outside and tough on the inside. The best way to maintain the sharpness of the blade is to regularly sharpen the cleaver. The **knife sharpeners** used to carry a load, **peddling** in a **high-pitched voice** through the streets. The knife sharpener industry gradually **faded out** of people's life. Instead, the majority of Chinese people use a **whetstone** or **sharpening stick**, even the back of the bowl sometimes, to sharpen their own knives at home.

Practicing how to use the Chinese cleaver formulates the skill of the Chinese chef. **Knife skill** is the essential starting point for any would-be chef. In China, **apprentices** studying in the culinary school receive a Chinese cleaver on their first day of training. Every day, they have to take the cleaver to class, and learn knife sharpening and knife skills. Every lesson would include instructions for chopping, slicing, and **shredding** of the raw ingredients with the Chinese cleaver and other knife skills. There are more than 100 kinds of knife skills with explicit names. Even for the most common fried kidney, more than ten kinds of knife skills are available: the kidney can be cut into the shape of wheat ears, rain cape, lychee, Chinese character of longevity, phoenix tail and so on. Wensi Tofu (Minced Tofu Soup), a famous Huaiyang cuisine in China, is more like a test for the chef than just presenting the taste. Shredding the soft tofu into hair-like pieces as if cutting a carrot is not only testing the cooperation of chef's hands, eyes and cleaver, but also the unity of his confidence and skill.

For a chef with excellent knife skills, his skills come second—the more important element is his inner peace. Using a cleaver skillfully has harsh requirements for his age. But for a successful chef, he relies not only on his youth vitality, but also on his profound experience by age.

Vocabulary

cleaver /ˈkliːvə(r)/ *n.* 菜刀

blade /bleɪd/ *n.* 刀刃；刀片

debone /diːˈbəʊn/ *v.* 将……去骨

flat /flæt/ *n.* 刀面

whack /wæk/ *n.* 重击

mince /mɪns/ *v.* 切碎

paste /peɪst/ *n.* 肉（或鱼等）酱

shovel up 铲起

slash /slæʃ/ *v.* 砍

dice /daɪs/ *v.* 切丁

carbon steel 碳钢

stainless steel 不锈钢

smash /smæʃ/ *v.* 猛击

quench /kwentʃ/ *v.* 淬火

knife sharpener 磨刀匠

peddle /ˈpedl/ *v.* 沿街叫卖

high-pitched voice 高音；调门高的嗓音

fade out 渐渐淡出；逐渐淡出

whetstone /ˈwetstəʊn/ *n.* 磨刀石

sharpening stick 磨刀棒

knife skill 刀工

apprentice /əˈprentɪs/ *n.* 学徒

shred /ʃred/ *v.* 切丝

Tasks

 I. Group Talk

In the ancient story "Ju An Qi Mei" (lifting the tray to the eyebrow level) indicating mutual respect in marriage, "An" refers to the small tray used in the past. Search for information about the tray and discuss the ancient dining style based on the use of trays.

 II. Writing

The story Dismember an Ox as Skillful as a Butcher in (Pao Ding Jie Niu) Chuang Tzu shows the butchers' superb cutting skills. Write an article of about 150 words to illustrate your insights into this story.

_____

_____

_____

_____

_____

_____

_____

_____

_____

_____

_____

_____

III. Translation

Translate the following sentences into English.

1. 中国自古以来就是一个文明礼仪之邦。这种"文明礼仪"表现在饮食文化上，便出现了诸多的宴席礼节。

   _____

   _____

2. 中国厨房里大部分的烹饪用一口炒锅、一个蒸笼、一个电饭锅或煮饭锅、一双筷子和一把瓢子就能搞定。

   _____

   _____

3. 菜刀要谨慎放置，不要放在可能受到强烈冲击或者引起事故的地方。

   _____

   _____

4. 中餐的刀工艺术最有趣的部分是其中的美学。中国厨师说起一道菜到底好不好，标准通常都是"色香味形"。

   _____

   _____

5. 川菜厨师们对刀工和火候的精妙掌握是闻名遐迩的。

   _____

   _____

IV. Short Video

Chinese chefs are experts of knife skills. Search for delicacies which show knife skills. Make a short video to illustrate how one of these dishes is done, and share it with your classmates.

Chinese Wok

There is one utensil you could not miss in the Chinese kitchen—the wok. The traditional Chinese wok is perfect for **stir-frying**: the curved bottom allows for even heating and for the easy movement of food around its hot bottom. Woks can also be used for **deep-frying**, boiling, steaming and **dry-roasting**, etc. Most Chinese woks are made from carbon steel, crude or cast iron. All of these are superior material for wok, but they must be **seasoned** to prevent rusting.

Before using a new wok, one needs to season it. It needs to be scoured thoroughly with wire wool to remove any rust or coating. After it is washed and dried, it is heated over a high fire. When the wok is extremely hot, it is carefully wiped with paper towels soaked in cooking oil. After the wok has cooled down, it is wiped or rinsed, and then get dried thoroughly.

Regular **re-seasoning** of a wok is quite simple, but it is an essential part of using the wok for stir-frying. Because it could create a classic Chinese wok with a non-stick surface, ensuring that food can be moved around easily. If food, especially meat, poultry and fish, is sticking to the wok, it's most likely the result of improper seasoning. To re-season the wok, one needs to heat it over high heat as the first time until it is extremely hot. Then add a few tablespoons of cooking oil and carefully **swirl** the wok gently. When the smoke is **rising** around the edges, pour the oil into a **heat-resistant** container. Then pour some fresh oil and heat it to the desired temperature for cooking.

Chinese chefs prefer a **ladle** for stir-frying because it is so **versatile**, while most Chinese people choose the wok scoop. The ladle can be used to stir food around the wok, ladle stock into sauces, transfer cooked food into serving bowls, and also to measure out ingredients and mix up last-minute sauces. Instead, **wok scoop** is used at home, which is better for stir-frying ingredients and scraping the wok bottom.

The wok is not only popular among the Chinese, but also well-known in the Western countries. Since Ken Hom[1] introduced the concept of the Chinese wok to the UK **from scratch**, Westerners started to understand the miracles brought by the Chinese wok. "On average, 12.5% households in the UK has a Ken Hom Chinese wok," said by Ken Hom in 2017. The wok is deservedly the most common kitchen utensil in the Chinese kitchen, for it can meet almost all cooking needs. The Chinese can make all kinds of delicious dishes with just two kitchen utensils: a Chinese cleaver and a wok.

Note

[1] Ken Hom: Ken Hom OBE (谭荣辉), (born May 3, 1949 in Tucson, Arizona, United States) with ancestry from Taishan, Guangdong, is a notable Chinese American chef, author and British television-show presenter. In 2009, he was awarded by HM, the Queen with an honorary OBE for "services to culinary arts". The OBE (Officer of the Most Excellent Order of the British Empire) recognizes his achievements and the impressive social and historical impact he has made on the way the UK has "adopted" Chinese cuisine, which has now become one of the UK's favorites.

Vocabulary

stir-frying /ˈstɜː(r)fraɪŋ/ *n.* 炒

deep-frying /ˈdiːpfraɪŋ/ *n.* 炸

dry-roasting /ˈdraɪrəʊstɪŋ/ *n.* 干烧

season /ˈsiːzn/ *v.* 开锅

re-season /riːˈsiːzn/ *v.* 养锅

swirl /swɜːl/ *v.* 打转；旋转

rinse /rɪns/ *v.* 冲洗

heat-resistant /ˈhiːtˌrɪzɪstənt/ *adj.* 耐热的

ladle /ˈleɪdl/ *n.* 瓢子

versatile /ˈvɜːsətaɪl/ *adj.* 通用的；万能的

wok scoop 锅铲

from scratch 从无到有；从头做起；白手起家

Tasks

I. Group Talk

Discuss common utensils found in Chinese and Western dishes and share with your classmates.

II. Writing

It was said that utensils outshined food (Mei Shi Bu Ru Mei Qi). Fine combination between food and utensils is a critical factor of the glamorous Chinese cooking art. Search for materials to write an article of about 150 words on key points of matching utensils and dishes.

_____

_____

_____

_____

_____

_____

_____

_____

_____

_____

_____

_____

III. Translation

Translate the following sentences into English.

1. 食在中国，味在四川。只要在蜀地吃过川菜，就知道这句话毫不夸张。

   _____

2. 四川小吃名扬天下（成都小吃是其中的一枝花）：水饺、抄手、面条和其他种种，鲜香美味，不一而足。

   _____

3. 腌制的蔬菜在川菜中占有一席之地，很多人还会自己做快手的"洗澡泡菜"，供日常吃食。

   _____

4. 有些炒锅有一个长长的锅柄，方便掂锅，但四川最常见的还是小小的金属耳柄。

   _____

5. 竹质锅刷，把细竹板捆起来，是在炒菜间隙做清洁炒锅之用。

   _____

   _____

IV. Short Video

Make a short video to explain how to use chopsticks, and share it with your classmates.

Supplementary Reading

German Kitchenware

Located in the middle of Europe, Germany is one of the most economically developed countries in the world and belongs to the European food culture circle. Although German food is not as rich and diverse as French food, nor as world-renowned as Italian food, it has developed its own unique food culture through history. Germany's high latitude and temperate monsoon climate are characterized by long winters and short unfrozen periods, which retards crop production. However, pastures in the north are abundant and conducive to animal husbandry, which explains why Germans consume a large amount of sausages. The relatively cold climate is suitable for growing hops, so beer is more popular in Germany than in other European countries. Schweinshaxe (pork knuckle) and Sauerkraut (sour cabbage) are also distinctive dishes on German tables.

When you walk into the kitchen of a German home, you will feel bewildered by the variety of kitchenware. In Germany, almost every family has a full set of pots and pans for frying meat, cooking vegetables, making pasta, and making steak. The division of kitchenware usage is extremely detailed. Some people once raved that this is not a kitchen, but a laboratory with pots, pans, knives, and measuring cups.

Why do Germans have such high requirements for kitchenware? The kitchen has a vital meaning to them. It is not only a place for cooking, but also a bridge to maintain the relationships among family and friends. Besides the living room, the kitchen is often the most common place for German people to meet and talk. If young people invite friends at night, they will often bring them to the kitchen to eat something and have a night chat. In addition to the three meals a day, some Germans are also accustomed to a four or five o'clock "extra meal", which includes a cup of coffee or tea, a piece of cake, or a few cookies. Many families would invite friends home to drink tea and have a nice talk at that time. The kitchen plays the role of a meeting room. Therefore, the requirements for kitchenware consumption are naturally higher.

The high requirements for kitchenware are also reflected in the pursuit of high quality. Take a closer look at German kitchenware, you will find that knives, pots, and pans are mostly made of 18/10 stainless steel, a material used in the manufacture of surgical knives, which is resistant to high temperatures and strong acids and bases. The material enables German kitchenware to be durable. The Germans seem to take the kitchen as a test site for various new materials, and the strict selection of materials has long been a tradition in German design. For example, the famous German product and furniture design studio — Formfürsorge Studio has shown its keen interest in experimenting with materials for kitchenware. They have made cutting boards for the kitchen out of non-slip material originally used in the manufacture of conveyor belts and polyethylene.

The experimental exploration of materials reflects the inquisitive and creative spirit of the Germans, just as the exploration of modernist design style in the early 20th century did. The Germans maintained this functionalist style belief in design once they felt best suited their spirituality. Germans apply their rigor and precision to the design of kitchenware, thus creating a series of world-renowned kitchenware brands. Even on a global scale, products made in Germany are always popular for their robustness and durability which is also a reflection of the Germanic national character.

Assorted Pickles

Ingredients

500g varieties of vegetables; 200g old pickle brine; 800g cold boiled water; 5g Baijiu alcohol; 10g ginger, sliced; 40g salt; 10g dry chilies, cut into sections; 2g Sichuan peppercorns; 3g star anise; 3g cinnamon; 3g amomum tsao-ko; 15g rock sugar

Preparation

1. Remove any rough roots, withered leaves, and rough skin of the vegetables. Clean thoroughly, cut into strips, and drain.

2. Mix old pickle brine with cold boiled water, alcohol, ginger, salt, rock sugar, star anise, cinnamon, amomum tsao-ko, dry chilies and Sichuan peppercorn in an earthen container used exclusively for Sichuan pickles to get new pickle brine.

3. Add varieties of vegetables, cover and seal the container, leaving all ingredients soaked for 5 hours. Remove the vegetables to a serving dish, and season with chili oil, MSG and sugar to taste.

什锦泡菜

食材配方

各种蔬菜 500 克、老泡菜盐 200 克、凉开水 800 克、白酒 5 克、姜片 10 克、食盐 40 克、干辣椒 10 克、花椒 2 克、八角 3 克、桂皮 3 克、草果 3 克、冰糖 15 克

制作工艺

1. 各种蔬菜去老根、黄叶、粗皮，洗净，切成条，晾干水分。

2. 泡菜坛中加入老泡菜盐水、凉开水、白酒、姜片、食盐、冰糖、八角、桂皮、草果、干辣椒、花椒混合，调成新的泡菜盐水。

3. 将各种蔬菜放入泡菜坛，加盖密封，浸泡 5 小时后取出装盘，可根据需要酌加辣椒油、味精、白糖拌匀即成。

German Pork Knuckle

Ingredients

1,200g pork knuckle (hind knuckle, uncured); 1 bunch of soup vegetables; 2 cloves of garlic; 1 onion; 1 bottle of stout; 3 bay leaves; 3 cloves; 3—4 juniper berries; 3—5 caraway seeds (depending on your taste); 10 peppercorns; a few coarsely ground pepper; a little salt; a little cumin powder

Preparation

1. Wash the pork knuckle and put it in a soup pot with cold water. Add a spoonful of salt to the water. Dice the soup vegetables, crush the garlic, and cut the onion into quarters. Prepare spices: 3 cloves, 3 bay leaves, 3—4 juniper berries, caraway seeds as needed and about 10 peppercorns. Put spices in a tea infuser or a tea bag.

2. When the water boils, turn to low heat. Add soup vegetables, garlic, onions and spices. Simmer for 1.5 to 2 hours. Towards the end of the cooking time, preheat the oven to 200 °C top/bottom heat.

3. Take the knuckle out of the broth and use a knife to cut the rind in a diamond pattern. (Be careful not to damage the meat). Rub the rind with salt, coarsely ground pepper and cumin powder.

4. Place the knuckle on a griddle and put it in the oven. Bake for 45 minutes at 200 °C top/bottom heat and brush the rind regularly with the stout (stout can also be replaced with salt water). Then bake on top heat for 15 minutes at 220 °C while keeping an eye on the knuckle to prevent the crispy from turning black.

5. Place the knuckle on a plate and serve with coleslaw, mustard and bread.

德国烤猪肘

食材配方

猪肘（后肘，未腌制）1200 克、汤菜 1 份、大蒜 2 瓣、洋葱 1 个、黑啤 1 瓶、月桂叶 3 片、丁香 3 个、杜松子 3 ~ 4 个、香菜籽 3 ~ 5 粒（依据个人口味适当添加）、胡椒 10 粒、粗磨胡椒粉少许、盐少许、孜然粉少许

制作工艺

1. 将猪肘洗净，放入装有冷水的汤锅中，加入一勺盐。将汤菜切丁，大蒜压碎，洋葱切成四瓣。准备香料：3 个丁香、3 片月桂叶、3 ~ 4 个杜松子、香菜籽（依据个人口味适当添加）和约 10 粒胡椒。将香料放入泡茶器或茶包。

2. 水烧开后，转小火。加入汤菜、大蒜、洋葱和香料，炖煮 1.5 到 2 小时。在时间快结束时，提前将烤箱预热至上火 / 下火位 200℃。

3. 将猪肘从肉汤中取出，用刀将外皮切成菱形图案（注意不要切到肉）。用盐、粗磨胡椒粉和孜然粉涂抹外皮。

4. 将猪肘放在网格烤盘上，然后放入烤箱。上火 / 下火位 200℃烘烤 45 分钟，并定期用黑啤涂抹外皮（黑啤也可用盐水代替），然后上火位 220℃烘烤 15 分钟，同时注意观察，猪肘烤至酥脆为止，以防烤糊。

5. 将猪肘放在盘子上，与凉拌卷心菜、芥末和面包一起食用。

CHAPTER V

INGREDIENTS

Freshness of Ingredients

Passage Ⓐ

In China, people prefer going to local wet markets to buy fresh vegetables and meat on a daily basis while American people might choose **groceries** or supermarkets on a weekly basis. It is quite popular in China to see market **stalls** selling live fish, chicken or rabbit, where the **vendors** would slaughter, rinse and pre-process whichever the customer has chosen. But Westerners may not be accustomed to this because they think it is quite scary or brutal to watch killings in front of them. As for the Chinese, it could be attributed to a traditional idea: freshness.

Freshness in Chinese is composed of **radicals**: fish and goat. The character was first found on **bronzeware** dating from 3,000 years ago. At the beginning, it had a **pictorial** form consisting of a "goat" radical on the top and a "fish" radical below. Gradually, it had changed into the left and right form nowadays. In Chinese culture, the goat radical stood for the meaning "delicious". The story behind this Chinese character goes like this. One day, Xi Yu, the little son of Ancient Peng (Peng Zu, a legendary long-lived figure in ancient China) went back home with fish and asked his mom to cook. Then his mom chopped down some mutton at hand and cooked it with the fish together. When Ancient Peng came back and had the soup for dinner, he was amazed by its fresh taste. His wife told him about the preparation of the dish. After this, Ancient Peng tried several times by cooking fish and mutton together and had the same delicious soup as well. Hence, in Chinese, we had fish and goat radicals together standing for freshness.

Chinese consumers regard freshness as one of the most important aspect when they shop for food. A good Chinese chef insists on using extremely fresh ingredients, usually from the local market. Unlike Western countries, Chinese often go to local markets to buy fresh ingredients. Local markets, wet markets or "Caishichang" in Chinese, is a "traditional" form of food retail. Although China has a rapid **expansion** of "modern" supermarket chains, local markets have maintained their popularity in urban China. Their continued popularity rests in the freshness of their food. Independent vendors primarily sell "wet" items such as meat, **poultry**, seafood, vegetables and fruits. Wet markets not only **thrive** in China, but also continue to be important food providers in Malaysia, Indonesia, the Philippines, and Vietnam, as well as in more economically developed regions such as Singapore.

Although we have identified fresh food as the key to consumers in China, freshness is an **abstract** idea which is a more or less self-explanatory quality viewed by vendors and consumers. There is no certain **norm** to identify freshness in food. Taking cucumbers as an example, some people would view a yellow flower **attaching** at the bottom end or a **prickling** feeling by hands to identify the freshness of cucumber. As for poultry and fish, traditionally, people prefer to have chicken or fish killed on the spot to identify its freshness. Some dishes have a high requirement for fresh ingredients, for example, Braised Paddy Eels. The slaughtered and pre-processed paddy eels need to be cooked in an hour, otherwise their flavor **spoils**.

For fine cuisine and beverages, fresh-harvested ingredients are essential. What's your way to identify the freshness of ingredients?

Vocabulary

grocery /ˈɡrəʊsəri/ *n.* 食品杂货店

stall /stɔːl/ *n.* 货摊

vendor /ˈvendə(r)/ *n.* （尤指街上的）小贩

radical /ˈrædɪkl/ *n.* （汉字）偏旁；部首

bronzeware /ˈbrɒnzweə(r)/ *n.* 青铜器

pictorial /pɪkˈtɔːriəl/ *adj.* 图画的；照片的

expansion /ɪkˈspænʃn/ *n.* 扩大；增加

poultry /ˈpəʊltri/ *n.* 家禽

thrive /θraɪv/ *v.* 兴旺；欣欣向荣

abstract /ˈæbstrækt/ *adj.* 抽象的

norm /nɔːm/ *n.* 常态；标准；准则

attach /əˈtætʃ/ *v.* 连接；附上

prickling /ˈprɪklɪŋ/ *adj.* 刺痛的

spoil /spɔɪl/ *v.* （开始）变质；变坏

Tasks

 I. Group Talk

Freshness is a key part when Chinese people buy food ingredients. Can you think of more examples? Talk with your partners and share your opinions.

 II. Writing

How do your family members choose fresh food ingredients? Write an article of no less than 80 words.

_____

_____

_____

_____

_____

_____

_____

_____

_____

_____

_____

_____

_____

_____

III. Translation

Translate the following sentences into English.

1. 常用的烹饪原料可以分成四大类，即动物类原料、植物类原料、干货类原料和调味料原料。

_____

_____

2. 质量上等的鲜猪肉，肌肉有光泽，色红均匀，脂肪洁白，外表微干，不粘手。

_____

_____

3. 猪肉是一种重要的食物，是蛋白质、脂肪、维生素以及矿物质的优质来源。

_____

_____

4. 菜肴可由单一的原料组成，也可由两种以上的原料组成。

_____

_____

5. 不同的食物原料有不同的质感，如豆腐的质感比较嫩，土豆的质感比较软，竹笋的质感比较脆。

_____

_____

IV. Short Video

Make a short video about the freshness of ingredients in your local cuisine, and share it with your classmates.

Passage B

Foods as Home Remedy

Food, at its most basic level, is fuel. It powers you, giving you the **oomph** that you need to make it through the day. But if you choose wisely, what you eat can also be healing. It is "Shiliao", also called food **therapy** or food cure. It is one of the **therapeutic** modalities in traditional Chinese medicine, which can be traced back for thousands of years in China. This science regards the human beings and other things in nature as combination of the two opposites, "Yin" and "Yang". The loss of balance between Yin and Yang gives rise to diseases, and to treat a disease is to readjust Yin and Yang. *Neijing*, the *Yellow Emperor's Classics of Internal Medicine* says, "If there is heat, cool it; if there is cold, warm it; if there is dryness, **moisten** it; if there is dampness, dry it; if there is **vacuity**, supplement it; and if there is excess, **drain** it." Here, cold, dampness and vacuity are Yin; heat, dryness and excess are Yang. With a Yin-Yang balance, the healing begins.

Likewise, whenever there is too much Yin or too much Yang in the food we consume, there are imbalances in the body which would produce illnesses. The Chinese believe that Yang foods increase the body's heat thereby raising the rate of **metabolism**, while the Yin foods lower the body's heat as well as the rate of metabolism. Generally, Yang foods have dense food energy while Yin foods have high water content. An ideal **macrobiotic** diet should have a proper balance of the Yin and the Yang for overall health and wellness. For instance, consuming too many Yang food items will cause **acne** and bad breath and too many Yin food items might **render** a person **anemic** or **lethargic**.

For example, when you have a cough and a sore throat, it means there is dryness in your throat and lungs, and you'd better drink water to moisturize them. Sometimes, elder people prefer using a traditional recipe — Steamed Pears with Rock Sugar to get a better medicinal effect. This dish uses pears and rock sugar as the main ingredients, and sometimes goes with Goji berries (wolfberries), Yin'er (tremella mushroom), Chenpi (dried tangerine peel) or Lianzi (lotus seeds). It is believed that steamed pears with rock sugar help to reduce **phlegm**, clear heat, moisten the lung and cool the heart. This dish is also good for dry cough or dry throat. It is amazing that a common fruit pear can have such magic effect on human body.

According to macrobiotics principles, there is really no complete Yang or complete Yin food. Everything is relative, just like in real life where nobody is totally good or bad. So, a dish prepared can be more Yin compared to another, but Yang compared to yet another dish. But there are foods that are **predominantly** Yang or Yin. Some examples of Yang or **contractive** food are salt, sea vegetables and seeds. Notice how you can envision what this food will do when you eat it. Salt will remove water from the cells and make it contract, which is Yang. Seeds are life forces tightly bound together, and thus very contractive, again Yang. Some examples of Yin or expansive foods are sugar, water and fruits.

Vocabulary

oomph /ʊmf/ *n.* 活力

therapy /ˈθerəpi/ *n.* 治疗；疗法

therapeutic /ˌθerəˈpjuːtɪk/ *adj.* 使人镇静的；使人放松的

moisten /ˈmɔɪsn/ *v.* 使湿润

vacuity /vəˈkjuːəti/ *n.* 虚；贫乏

drain /dreɪn/ *v.* 使排出

metabolism /məˈtæbəlɪzəm/ *n.* 新陈代谢

macrobiotic /ˌmækrəʊbaɪˈɒtɪk/ *adj.* 养生饮食的；延年益寿的

acne /ˈækni/ *n.* 痤疮；粉刺

render /ˈrendə(r)/ *v.* 使成为；使变得；使处于

anemic /əˈniːmɪk/ *adj.* 贫血的

lethargic /ləˈθɑːdʒɪk/ *adj.* 无精打采的；懒
洋洋的

phlegm /flem/ *n.* 痰；黏液

predominantly /prɪˈdɒmɪnəntli/ *adv.* 占主导
地位地；显著地

contractive /kənˈtræktɪv/ *adj.* 收缩的；有
收缩性的

Tasks

 I. Group Talk

Food remedies are common in our life. Have you tried some food remedies in your life? Choose one and try to describe it and share with your partners.

 II. Writing

Chinese people have believed that the balance of food could help human body prevent from disease since ancient times. How do you think about it? Do your research and write an article of no less than 80 words.

_____

_____

_____

_____

_____

_____

_____

_____

_____

_____

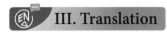 **III. Translation**

Translate the following sentences into English.

1. 古人把食物分为五色，相应的食物会对五脏起到补益作用。

   _____

   _____

2. 古人认为绿色对应木，肝为木，因此绿色食物有保护肝脏、滋养肝脏的作用。

   _____

   _____

3. 古人认为红色对应火，心为火，因此红色食物对心脏最有益。

   _____

   _____

4. 日常蔬果当中，大蒜、洋葱、芋头、豆芽菜，还有梨都是白色食物，有益于肺的保养。

   _____

   _____

5. 四季节气不同，我们在食疗的时候如果能顺应时节，那么日常饮食自然就能起到食疗的作用。

   _____

   _____

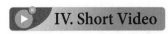 **IV. Short Video**

Make a short video about a Chinese food remedy, and share it with your classmates.

Sichuan Food Ingredients

Passage **C**

   Sichuan has long been known in China as the "Land of Abundance" (Tianfu Zhi Guo) because of its agricultural wealth. The **fertile** land of the Sichuan Basin, mild climate, rich rainfall and river-water resources create ideal conditions for farming. Around 2,300 years ago, the governor of Shu **Prefecture**, Li Bing, supervised the Dujiangyan **Irrigation** Project to **harness** the Minjiang River near Chengdu. It is an extraordinary construction which brought an end to the flooding that had **plagued** the region, opening the land for stable and productive agriculture, and ensuring that its people would always be well-fed.

   Fertile Sichuan land produces not only rice but all kinds of fruit and vegetables, all year around. Local specialties include bamboo shoots, celery, eggplants, lotus root, water spinach, Chinese chives, **gourds** of all shapes and sizes, mandarin oranges, **pomelos**, lychees, longans, peaches and loquats. Some other specialties are fine tea, wild vegetables such as **fiddlehead ferns** (Juecai) and the spring shoots of the Chinese toon tree (Chunya).

   Sichuan is **crisscrossed** by rivers and streams with rich types of fish, including the rock carp (Liyu), Ya'an snowtrout (Yayu) and long-snout catfish (Jiangtuan). The surrounding mountains, forests and grasslands once teemed with wildlife and exotic plants, such as **fungi**, wild frogs and all kinds of medicinal roots and herbs. Some people even describe Sichuan region as a vast **larder** filled with the stuff of **gastronomic** dreams.

Chili is a quite important part in Sichuan cuisine. The Sichuanese traditionally favor a couple of specific varieties. Most distinctive is the Erjingtiao, a long, thin-skinned, mild chili with a curved tail and an **enticing** flavor. It has a stronger taste compared with **plump**, round chilies known as "lantern chilies" (Denglongjiao), which are used in pickling or dish cooking. Other local varieties include the smaller, plumper "facing heaven" chili (Chaotianjiao) and the **squat**, pointy "bullet" chili (Zidantou). In recent years, local chefs increasingly rely on small, pointy chilies known as "little rice" chilies (Xiaomila), which are largely grown in other regions, including Yunnan.

Another key ingredient in Sichuan is Sichuan pepper. It has been grown in Sichuan since ancient times. The **spiky** plant, with green **pimpled** berries that grow rosy in the late-summer sun, prefers mountainous **terrain** with bright sunlight and freezing winters. Classic dishes of Sichuan cuisine wouldn't be the same without Sichuan peppercorns. When this fragrant but mouth-numbing spice is married with chili peppers, chefs believe this numbing effect reduces the chili pepper's heat, leaving diners free to appreciate the chili's intense, fruity flavor. Sichuan peppercorns can be used whole or ground into powder. The spice is one of the five ingredients that **comprise** five-spice powder (the others are star anise, fennel, clove and cinnamon), and it's used in many savory Sichuan dishes. Ground, roasted Sichuan peppercorn is used to make an **infused** Sichuan peppercorn oil. It is also paired with salt to make a flavorful Sichuan pepper salt to serve as a condiment with meat dishes.

Pickled vegetables also play a crucial role in local cooking, and many people still make their own quick-pickled vegetables for daily use; other pickles, which require more complicated processing, are usually bought. The most famous is Zhacai, which is peculiar to China. It enjoys the same fame as European pickles and Japanese tsukemono in the international market. Being one of the "four great pickles", it is made of a spiced, salted mustard **tuber** produced in Fuling, Chongqing. The others are "big-headed vegetable" (Datoucai), a kind of salted **turnip**; Nanchong "winter vegetable" (Dongcai), made from a leafy variety of **mustard**; and Yibin yacai, a dark, sweet, spiced preserve made from the tender stems of another mustard variety.

Vocabulary

fertile /ˈfɜːtaɪl/ *adj.* 肥沃的；富饶的

prefecture /ˈpriːfektʃə(r)/ *n.* 县；辖区；地方官

irrigation /ˌɪrɪˈgeɪʃn/ *n.* 灌溉

harness /ˈhɑːnɪs/ *v.* 控制（利用）（自然力等）

plague /pleɪg/ *v.* 困扰；折磨

gourd /gʊəd/ *n.* 葫芦类属植物

pomelo /ˈpɒmələʊ/ *n.* 柚子

fiddlehead fern 像小提琴头的蕨类植物

crisscross /ˈkrɪskrɒs/ *v.* 纵横交错

fungus /ˈfʌŋɡəs/ *n.* 菌类（复数为 fungi）

larder /ˈlɑːdə(r)/ *n.* （家中的）食品贮藏室

gastronomic /ˌɡæstrəˈnɑːmɪk/ *adj.* 美食的

entice /ɪnˈtaɪs/ *v.* 诱惑

plump /plʌmp/ *adj.* 丰满的；胖乎乎的

squat /skwɒt/ *adj.* 矮胖的；粗矮的

spiky /ˈspaɪki/ *adj.* 竖起的；带尖刺的

pimple /ˈpɪmpl/ *n.* （尤指长在脸上的）粉刺；小脓包

terrain /təˈreɪn/ *n.* 地形；地势

comprise /kəmˈpraɪz/ *v.* 包括；由……构成

infuse /ɪnˈfjuːz/ *v.* （辣椒面、茶叶等）被浸渍；泡；沏

tuber /ˈtjuːbə(r)/ *n.* （马铃薯等植物的）块茎

turnip /ˈtɜːnɪp/ *n.* 白萝卜；芜菁

mustard /ˈmʌstəd/ *n.* 芥菜

Tasks

 I. Group Talk

Which Sichuan ingredient impresses you most? Try to describe it and share with your partners.

 II. Writing

Sichuan cuisine has a wide selection of ingredients. Simple ingredients can be cooked with extraordinary flavors. Do you know any traditional Sichuan recipe that uses average ingredients? Write an article of about 80 words to introduce the recipe.

_____

_____

_____

_____

_____

_____

III. Translation

Translate the following sentences into English.

1. 川菜取材讲究时令。

_____

_____

2. 立春后的蒜薹，质地细嫩、鲜而少香，烹饪时则宜鸡丝辅之。

_____

_____

3. 清明前的蒜薹，其香、鲜、嫩兼之，佐家常味最适。

_____

_____

4. 谷雨后的蒜薹大批应市，其质地较老，本味浓郁，烹制上一般就以保鲜、抑香为主。

_____

_____

5. 莲花白春清脆而带异味，冬细嫩而味甜鲜。

_____

_____

IV. Short Video

Make a short video about a peculiar ingredient of Sichuan cuisine, and share it with your classmates.

Supplementary Reading

Butter

Oil is one of the most essential ingredients in cooking. In Chinese families, vegetable oil is commonly used. In the southern part, people prefer using rapeseed oil, while people live in the northern part like using peanut oil or soybean oil. In Western countries, butter dominates the oil–consuming top list. It is widely used in burger grilling, pizza baking, steak or bacon frying. It can add a milky flavor to dishes which makes people have a stronger appetite. Then, we would ask, why do they like butter so much? Unlike the Western countries, in China, butter is usually sold in supermarkets instead of grocery stores. What makes this difference?

First, we need to know, what is butter?

Butter is a dairy product created from proteins and fats found in milk and cream. In the US, most butter is cow milk-based, but butter also comes from many other sources such as milk from sheep, goats, buffalo, and yaks. It's created by churning or shaking cream until it separates into solid and liquid parts called butterfat and buttermilk, respectively. The flavor butter adds to sauces, breads and pastries is unequaled. Butter contains at least 80% milkfat, not more than 16% water and 2% to 4% milk solids. It may or may not contain added salt. Butter is firm when chilled and soft at room temperature. It melts into a liquid at approximately 33 °C and reaches the smoke point at 127 °C.

There are many different types of butter to try, like salted butter, ghee or clarified butter, whipped butter, spreadable butter, light butter, plant-based butter, and brown butter, etc. Besides milkfat, butter is a source of Vitamin A, Vitamin D, Vitamin E and calcium, which is beneficial to human body. Perhaps the most obvious role butter plays in baking is adding flavor to baked goods. The flavor butter adds to pastries, cakes, cookies, and more just really can't be mimicked. There are products that are "butter flavored" such as butter flavored shortening, but the richness that comes from real butter is distinct.

Then, we probably want to ask, how did people get butter from milk?

Butter making originated in prehistoric time. The time and place when people first created butter is still a subject of ongoing debate. Most anthropologists agree that butter arrived on the scene with Neolithic people, the first of our Stone Age ancestors who succeeded at domesticating ruminants. In fact, we wouldn't recognize the world's earliest butters. For one thing, they were made from the milk of sheep, yak, and goats, not from cow's milk. Domesticated cattle came much later in man's conquest of various animals. Once these early families had milking animals under their control, the invention of different dairy products was an evolutionary next step.

We'll never know the exact details, but the probable scenario went something like this: A herdsman's milk formed thick butter flakes floating accidently during his moving from one pasture to another. We could guess, the herdsman may have cursed his lumpy liquid at first, but he was marveled by the rich taste of the butter kernels. Eventually, this ancestor and his clan also discovered that this tangy dairy fat was not only useful for eating but also for cooking, fuel, and medicine. Around the 1st century AD, butter was common throughout most of the developing world.

However, there were significant exceptions as well, such as the Mediterranean with olive oil as its ruling food fat. Meanwhile, on the other side of the ancient world, most early Chinese communities, except those in the far north, would rarely see, let alone taste, butter since dairying itself was an anomaly. Partly, it was attributed to our staple food which was vegetables and cereals. And cattle were mainly used to plough the croplands, not for milking. An old saying is often heard in China, "I serve my people with my head bowing like a willing ox." Cattle in Chinese culture carry an image of diligence and perseverance. As time goes by, China has a variety of cooking oil, such as lard, rapeseed oil, peanut oil and so on. There is no need to use butter as cooking oil. As a result, milky flavor is not the preference of Chinese people. Because of these reasons, China didn't develop a diary culture, hence the difference in oil consumption between China and Western countries.

Recipe Sharing

Lemon Flavor Pastry Cake

Ingredients

12 lard short dough, 1 lemon, 50g lard, 50g sugar, 30g winter melon candy bars, 30g crispy peanuts, 1 egg yolk, 15g roasted sesames, 40g cooked wheat flour

Preparation

1. Chop lard, winter melon and crispy peanuts; squeeze out lemon juice.

2. Blend lard, lemon juice, sugar, winter melon candy bars, sugar, peanuts, sesames and fried wheat flour to make lemon stuffing.

3. Roll and flat the dough, put the lemon stuffing on it, fold into half, brush with whisked egg yolk. Put on the baking plate and bake for 20 minutes in a preheated oven. Then take out and place them on a serving plate.

柠檬酥

食材配方

千层油皮 12 张、柠檬 1 个、猪板油 50 克、白糖 50 克、蜜瓜条 30 克、酥花生 30 克、鸡蛋黄 1 个、熟芝麻 15 克、熟面粉 40 克

制作工艺

1. 猪板油、蜜瓜条、花生分别剁细；柠檬取汁。

2. 猪油、柠檬汁、白糖、蜜瓜条、花生米、芝麻、熟面粉拌成柠檬馅。

3. 将柠檬馅放在千层油皮上，对折成半圆形，刷上蛋黄液，装入烤盘，入烤箱烤制 20 分钟，取出后装盘。

Recipe Sharing

Butterscotch Pudding

6 servings, about 1/2 cup each

Ingredients

1¾ cups of whole milk, 1 cup of heavy cream, 1/4 cup of cornstarch, 3 large egg yolks, 1/2 teaspoon of salt, 6 tablespoons (3/4 stick) of unsalted butter, 1 cup of packed dark brown sugar, 2 teaspoons of vanilla extract

Preparation

1. In a pitcher, combine the milk, cream, cornstarch, egg yolks, and salt. Whisk until combined and set aside.

2. In a medium saucepan over medium heat, melt the butter. Add the sugar and reduce the heat to low. Stirring frequently, let the mixture cook for about 2 minutes.

3. Gradually whisk the milk mixture into sugar mixture in a thin steady stream. Increase the heat to medium and cook, stirring constantly, until it begins to bubble and thicken. Let it cook for another minute, then remove from the heat.

4. Stir in the vanilla and divide among glasses or pudding cups. Cool until warm, then cover the glasses with plastic wrap (keep the plastic from touching the surface of the butterscotch) and chill in the fridge for 1 to 2 hours, until set.

司考奇布丁

制作 6 份，每份约 1/2 杯

食材配方

1 杯全脂牛奶、1 杯浓奶油、 1/4 杯玉米淀粉、3 个大蛋黄、1/2 茶匙盐、6 汤匙（3/4 条）无盐黄油、1 杯深色红糖、2 茶匙香草精

制作工艺

1. 将牛奶、奶油、玉米淀粉、蛋黄和盐倒入一个罐子，搅打混合。放置备用。

2. 将黄油放入一个中号炖锅，用中火熔化，加入糖，转成小火。不断搅拌，煮 2 分钟左右。

3. 匀速向锅中逐步倒入牛奶混合液，边倒边搅拌。转成中火，持续搅拌直到液体开始变稠冒泡。再多煮 1 分钟，然后锅离火。

4. 加入香草精搅拌均匀，把液体倒入几个玻璃杯或布丁杯中。晾至温热，然后用保鲜膜盖住杯口（不要让保鲜膜接触到液体表面），放入冰箱冷藏 1~2 小时直到定型。

SEASONINGS AND SPICES

Sichuan Traditional Seasonings

Sichuan Province of China is famous for not only its beautiful scenery, but also its delicious cuisine. Speaking of the cuisine, the first word that springs to mind might be spicy. It is often heard that no trip to Sichuan would be complete without a bite of spicy food. So what makes Sichuan cuisine spicy so much? Is the spicy flavor different from that of other provinces? The story may unfold with the climate of Sichuan in the first place.

Sichuan is a **landlocked** province in southwest China and occupies most of the Sichuan Basin and the easternmost of Tibetan Plateau. Sichuan boasts a mild and **humid** climate, and sunny days are rarely seen as it is rainy or cloudy almost all the year round. According to the principles of traditional Chinese medicine, humidity leads to **accumulation** of **dampness** in the body which **translates** into diseases like **rheumatism** and **lethargy**. To adapt to the climate, smart Sihcuan people have been cooking with their spicy seasonings since the ancient times to remove excessive water in their body. When it comes to the various Sichuan seasonings, bean paste, chili pepper and Sichuan pepper are the three most important chapters one should not just skim.

Bean paste is one of the most commonly used seasonings in the kitchens of the Sichuanese and those people have never stopped making and improving the paste since the Qing Dynasty. Of all brands of bean paste in Sichuan, Pixian Bean Past, produced in Pixian County (now Pidu District), Chengdu City, ranks first in the locals' heart and wins the reputation of the "Soul of Sichuan Cuisine" due to its amazingly good taste. It is no secret now that chili pepper and broad bean are the main ingredients to make dishes yummy. After the historically-inherited fermentation which lasts one year at least, the mixture turns out to be **reddish**, spicy and thick paste. Pixian Bean Paste has dominated the culinary culture of Sichuan for over three hundred years.

Chili pepper, another popular Sichuan seasoning, takes up nearly a half of the shelves in seasoning section of supermarkets. So by taking a quick glimpse of the shelves, you will never doubt that Sichuan people love the spicy flavor crazily. Chili pepper is basically categorized as dried chili pepper, chili pepper powder and pickled chili pepper. For instance, "Erjingtiao" chili pepper[1] is usually adopted for making dried chili pepper. You could put fresh "Erjingtiao" chili pepper in the direct sunlight until the water in it is completely evaporated. Hence the pepper turned out be dried chili pepper. If you cut dried chili pepper into sections, and finely slice and grind them, you will get chili pepper powder. To make prickled chili pepper, you should, in the first place, soak the fresh red chili pepper in the specially made salt brine for a couple of days, and then the chili pepper becomes a salty and spicy condiment which can greatly whet people's appetite at the beginning of dinner.

Since the 21st century, biotechnology has taken a great leap forward across the world. A wide variety of hi-tech seasonings have been developed and brought into production such as chicken essence, savory salad seasoning and peanut butter. However, in Sichuan cuisine, traditional seasonings like Pixian Bean Paste can never be replaced by those newly invented ones. For centuries, those traditional seasonings have been not only **diversifying** the flavors of food, but also passing on the heritage of Sichuan culinary culture. They are never just something that make dishes delicious but are the fruits of diligence and intelligence of Sichuan people.

Note

[1] Erjingtiao chili pepper: It is a variety of chili that is most common in Sichuan cuisine of China. The chili is typically shaped like the letter "J" and is between five and six inches long. This chili is known for its deep color and robust fragrance, and is often used in chili oil for that reason.

Vocabulary

landlocked /ˈlændlɒkt/ *adj.* 内陆的

humid /ˈhjuːmɪd/ *adj.* 潮湿的

accumulation /əˌkjuːmjəˈleɪʃn/ *n.* 积累

dampness /ˈdæmpnəs/ *n.* 湿气

translate /trænsˈleɪt/ *v.* 转化

rheumatism /ˈruːmətɪzəm/ *n.* 风湿病

lethargy /ˈleθədʒi/ *n.* 无精打采

reddish /ˈredɪʃ/ *adj.* 微红的

diversify /daɪˈvɜːsɪfaɪ/ *v.* （使）多样化

Tasks

 ### I. Group Talk

With Sichuan traditional seasonings, cooks make dishes favored by people around the globe such as Kung Pao Chicken[1] and Mapo Tofu[2]. Besides, many other Sichuan dishes are worth introducing to foreign friends. Can you name one to your partners? And explain what seasoning you will use if you cook it yourself.

Notes

[1] Kung Pao Chicken, also transcribed Gong Bao or Kung Po, is a spicy, stir-fried Chinese dish made with cubes of chicken, peanuts, vegetables (traditionally Welsh onion only), and chili peppers.

[2] Mapo Tofu is a popular Chinese dish from Sichuan Province. It consists of tofu set in a spicy sauce, typically a thin, oily and bright red suspension, based on douban (fermented broad bean and chili paste) and douchi (fermented black beans), along with minced meat, traditionally beef.

 ### II. Writing

Erjingtiao chili pepper, a popular seasoning of Sichuan cuisine, is often seen on the table of the Sichuanese and it is indispensable in cooking many cold and hot dishes. Can you describe Erjingtiao chili pepper from different aspects like the place where it grows, shape, taste or other aspects you could come up with. Write an article of about 80 words.

_____

_____

_____

_____

III. Translation

Translate the following sentences into English.

1. 食无定味，适口者珍。

_____

_____

2. 郫县豆瓣是川味食谱中常用的调味品，有"川菜之魂"之称。

_____

_____

3. 川菜的魅力是独特的，总是能改变许多人的口味，把不吃辣的变得无辣不欢。

_____

_____

4. 辣椒中的辣椒素有降血压、降胆固醇的功效，从而能有效地保护人们免受心脏病的侵袭。

_____

_____

5. 调味时应注意根据季节的变化掌握口味，如天气热时味道就宜清淡些，天气冷时味道就宜浓重些。

_____

_____

IV. Short Video

Make a short video about one Sichuan seasoning, and try to introduce it to your classmates.

The Soul of Sichuan Cuisine

Pixian Bean Paste, produced in Pixian County (now Pidu District), Chengdu City, Sichuan Province, is one of the most famous seasonings of Sichuan cuisine. This household name enjoys a history of three hundred years and has ranked among the top Chinese condiments. The brightly thick paste now has won the **reputation** of the "Soul of Sichuan Cuisine", **indispensable** for preparing Sichuan dishes. The technique of producing bean paste has been listed on the Second Catalogue of National Intangible Cultural Heritage in China, which **demonstrates** that the paste is favored by not only the Sichuanese, but also people from other parts across China.

According to local legends of Pixian County, in Qing Dynasty, the Chens, a large family from southeast China, were migrating to Sichuan Province, carrying the cooked broad beans with them. After the Chens entered Sichuan, the beans gradually turned **moldy** due to the humid climate. Considering the shortage of food, Chen did not throw the moldy beans away. He spread them on the rocks, mixed them with chili pepper and salt, and had a bite. To his surprise, the thick **mixture** tasted not bad. It was refreshing with a spicy flavor. After settling down in Pixian County, Chen still couldn't forget the mixture's flavor at all and tried to remake it again. Through several days of working, he was excited to enjoy the delicate flavor again. And that was where the **legend** of Pixian Bean Paste began.

During the past three hundred years, the technique of producing Pixian Bean Paste has been constantly improved. The paste now has been under the geographical indication protection in China with its quality assessment standards published in 2005. If you want to make bean paste yourself, you should prepare great ingredients for sure as well as an adequate amount of time. The main ingredients include Erjingtiao chili pepper, broad beans, salt and water. While everything is ready, you are supposed to adopt the following procedure: first, soak and de-shell the broad beans to make the **starter** culture and **ferment** the beans for over six months; second, salt and crush the Erjingtiao chili and ferment the chili for the chili culture; third, mix the starter culture with the chili culture and place the mixture in the sunlight for fermentation for another three months; and then the bean paste is finally yielded. With this amazing seasoning, the locals cook delicate dishes such as Huiguorou (Twice-Cooked Pork with Chili Seasoning) and Fenzhengrou (Steamed Pork Wrapped in Lotus Leaves).

Apart from providing dishes with a great local flavor, the bean paste contains rich nutrition that human body needs. The protein and vitamins it has can prevent arteriosclerosis, lower cholesterol, promote **intestinal peristalsis** and increase appetite. Besides, the paste is also rich in choline, which is great for human's brain and can greatly enhance memory.

With the passage of time, some traditions, customs or local specialty products have long gone in human history. However, it is a blessing that the recipe of Pixian Bean Paste has been handed down from the ancestors and the technique of making the paste improved from generation to generation. The story about chili pepper and broad beans have been told for three hundred years and the world now is expecting how Pixian people will continue the legend of the "Soul of Sichuan Cuisine".

Vocabulary

reputation /ˌrepjuˈteɪʃn/ n. 名声

indispensable /ˌɪndɪˈspensəbl/ adj. 不可或缺的

demonstrate /ˈdemənstreɪt/ v. 显示

moldy /ˈməʊldi/ adj. 发霉的

mixture /ˈmɪkstʃə(r)/ n. 混合物

legend /ˈledʒənd/ n. 传奇

starter /ˈstɑːtə(r)/ n. 起子培养物

ferment /fəˈment/ v. 发酵

intestinal /ɪnˈtestɪnl/ adj. 肠的

peristalsis /ˌperɪˈstælsɪs/ n. 蠕动

Tasks

 I. Group Talk

Pixian Bean Paste is known as the "Soul of Sichuan Cuisine", which shows its importance in Sichuan culinary culture. After reading the passage, how much do you know about Pixian Bean Paste? What else do you want to know about it? You may turn to the Internet for more stories about Pixian Bean Paste and share them with your partners.

 II. Writing

Pixian Bean Paste dominated Sichuan seasonings for hundreds of years and dishes seasoned with it are welcomed by people from different parts of the world. Some concern that although the paste is a good choice for seasoning, innovation will never be made in Sichuan cuisine if Pixian Bean Paste remains dominant. To what extent do you agree or disagree with the statement. Write an article of about 80 words to explain the reasons why you think so.

_____

_____

_____

_____

_____

_____

_____

_____

_____

_____

III. Translation

Translate the following sentences into English.

1. 郫县豆瓣是成都市郫都区的特产，也是中国地理标志产品。

_____

_____

2. 郫都区属盆地中亚热带湿润气候，自然条件得天独厚。

_____

_____

3. 郫县豆瓣自从诞生的那一天起，就与蜀中百姓和所有喜欢川菜的人结下了不解之缘。

_____

_____

4. 郫县豆瓣含有丰富的蛋白质、脂肪、碳水化合物和维生素 C 等营养元素，长期食用可增进食欲，促进人体血液循环，并且起到驱湿祛寒的作用。

_____

_____

5. 作为川菜必不可少的辅料，郫县豆瓣对一方的文化和餐饮习俗均产生了深远的影响。

_____

_____

IV. Short Video

Make an interesting video or search one on the Internet about Pixian Bean Paste. The video should tell the audience how to make the paste or the development of its manufacturing technique.

The Secret of Spiciness in Sichuan Cuisine

Passage C

If cuisine of every province of China has a **distinctive** flavor, that of Sichuan dishes must be spicy. Chengdu has thousands of hot pot restaurants, let alone those serving spicy snacks along the streets. Before you explore what makes spiciness the **backbone** of Sichuan **culinary** culture, another question needs to be asked in the first place — what makes Sichuan cuisine spicy? And it's easy to arrive at the answer: chili pepper.

In kitchens of many countries, chili pepper is a great choice to add pungent "heat" to dishes and the Sichuanese are its **sophisticated** consumers. People in Sichuan would love to have dishes spiced up with chili pepper for at least one meal per day. But interestingly, not all of them know the truth that chili pepper did not **originate** in Sichuan Province, not even in China. Archaeologists found that the **Maya** people living in Mexico already discovered the edibility of chili pepper in the 5th century. In the 15th century, Christopher Columbus, an Italian **explorer** and **navigator**, reached this **mysterious** land by sea, tasted the spicy red fruit and brought it back to Europe. Not until the late 16th century, was chili pepper introduced to China. Nowadays, this reddish fruit as a well-received spice has been adopted in all the eight schools of Chinese cuisine and widely used in Sichuan and Hunan cuisines.

Why do Sichuan people like spicy flavor so much? There are basically three reasons. First, the Sichuanese had grown a liking for pungency before chili pepper was introduced to Sichuan. In the Eastern Jin Dynasty, Sichuan people favored three ordinary pungent seasonings: Sichuan pepper, **cornel** and ginger. After chili pepper arrived in Sichuan, it did not take long for the locals to accept this lovely plant with a familiar pungent flavor. Second, the spiciness of chili pepper counteracts the humid climate of Sichuan Province. According to the traditional Chinese medicine, climatic humidity leads to accumulation of dampness in the body, which could further result in diseases like rheumatism and **arthritis**. Chili pepper could act upon the **spleen** channel, and clear the fluid **retention**, dampness and cold. Third, the cost of enjoying chili pepper is acceptable to ordinary families. In the Eastern Jin Dynasty, people found that chili pepper, compared with cornel, satisfied their demand for pungency with less labor force and money to process and preserve. Later, chili pepper even replaced cornel to be one of the three main seasonings in the kitchen of Sichuan people.

Chili pepper not only makes dishes more delicious, but also brings many benefits to people's health. Studies show capsaicin in chili pepper can fool the brain into thinking that the soft **tissue** is being burnt so as to trigger the body to release **endorphins**. Endorphins, the natural pain-killing chemicals, can produce a feeling of euphoria. This physiologically interprets why the Sichuanese would rather constantly wipe the sweat from their forehead than give up the bite of reddish spicy delicacy.

Over half a century ago, chili pepper came all the way to Sichuan to enrich the flavors of food there. Now it has become the **essence** of Sichuan culinary civilization. Even the redness of chili pepper is regarded as the representative color of Sichuan Province, and **symbolizes** the locals' passion for delicacy and enthusiasm for life. If you haven't enjoyed Sichuan cuisine yet, don't hesitate to give it a try. Maybe you will fall in love with it after the first bite.

Vocabulary

distinctive /dɪˈstɪŋktɪv/ *adj.* 独特的

backbone /ˈbækbəʊn/ *n.* 支柱

culinary /ˈkʌlɪnəri/ *adj.* 烹饪的

sophisticated /səˈfɪstɪkeɪtɪd/ *adj.* 富有经验的

originate /əˈrɪdʒɪneɪt/ *v.* 发源

Maya /ˈmeɪə/ *n.* 玛雅人

explorer /ɪkˈsplɔːrə(r)/ *n.* 探险家

navigator /ˈnævɪgeɪtə(r)/ *n.* 航海家

mysterious /mɪˈstɪəriəs/ *n.* 神秘的

cornel /ˈkɔːnəl/ *n.* 山茱萸

arthritis /ɑːˈθraɪtɪs/ *n.* 关节炎

spleen /spliːn/ *n.* 脾脏

retention /rɪˈtenʃn/ *n.* 保留

tissue /ˈtɪʃuː/ *n.* 组织

endorphin /enˈdɔːfɪn/ *n.* 内啡肽

essence /ˈesns/ *n.* 精华

symbolize /ˈsɪmbəlaɪz/ *v.* 象征

Tasks

 I. Group Talk

One ancient Chinese account declares that the "people of Sichuan uphold good flavor and they are fond of hot and spicy taste". Are you a fan of spiciness? State the reasons to your partners.

 II. Writing

Apart from the Sichuanese, are there any other people fond of spiciness? Why do they eat spicy food? What makes their food spicy? Try to answer those questions or tell their stories of spiciness in your own way with about 80 words.

_____

_____

_____

_____

_____

III. Translation

Translate the following sentences into English.

1. 辣椒喜温暖、怕霜冻、忌高温。

---

2. 在我国，辣椒是很重要的调味品，甚至没有它就无法下饭。

---

3. 四川气候潮湿且经久不散，当地居民为了祛除体内寒气，时间长了就养成了吃辣的习惯。

---

4. 当年哥伦布发现美洲大陆后，将辣椒带到了欧洲，随后由欧洲传到了非洲和亚洲。

---

5. 辣椒的红象征四川人民对美食的热情以及对美好生活的向往。

IV. Short Video

In Sichuan restaurants, what dishes do you often order? Are they spiced up with chili pepper? Try to document the dishes with your camera next time you have dinner in a local restaurant and introduce one spicy dish to the audience.

Supplementary Reading

Spicy Culture in Mexico

Apart from people in Sichuan and Hunan who show profound love for spiciness, some in other countries cannot miss spicy food every single day. And Mexicans are among them. Mexico, a South American country, produces a variety of chili peppers and some chili pepper sauce used for cooking local cuisine is deliciously spicy, which is favored by foodies around the world.

Mexico is one of the origins of chili pepper. In 6000 BC, the Indians in South Mexico, Central America and Northern South America started to grow and eat chili peppers. Now, there are about 60 edible chili peppers cultivated in Mexico and over ten kinds of them are popular in local wet markets. The reason why people are so attracted to them is related to the geographic features of Mexico. The middle and southern Mexico boasts high latitude, which causes the locals sensory numbness to different extent. In order to mitigate the numbness, Mexicans, through hundreds of years of life experiences, found chili peppers could be a remedy to relieve such symptoms. Gradually, chili peppers have become an indispensable ingredient of Mexican cuisine and even been regarded as the soul of Mexican food.

Many classic Mexican dishes are spiced up with chili peppers. One of the most famous Mexican dishes is enchilada. It is a kind of meat roll with spicy sauce made from local chilies. Enchilada is so well-received a dish that it's almost seen on the menu of all Mexican restaurants around the globe. And its cooking is simple: A chef wraps up the pre-cooked chicken or beef with corn rolls and tops them with cheese and chili sauce. It can be made within two minutes and will definitely amaze you with its spicy and milky flavor. To cater to different tastes, the chef may choose different spicy sauces or stuff the rolls with different fillings, so there are possibly over a dozen of flavor combinations for customers to taste.

Another Mexico snack one should not miss is Chile Toreado. It's said that, before cooking, a chef rubs chili peppers thoroughly to "waken" the chili seeds inside them, so as to make the dish much spicier. Among all kinds of chili peppers, the chef prefers chile jalapeño or chile serrano to others. The chef fries the chili peppers with a little oil till slightly burned black. Sometimes, shredded onions are fried together with chilies. Before serving the dish, the chef will season the chili peppers with moderate salt and lemon juice. Chile Toreado tastes spicy as well as fresh and it going with wine is most favored by the locals.

Chile jalapeño mentioned above is one of the classic ingredients for Mexican cuisine. Over recent years, chili is more and more favored by people coming from European and American countries. With its large size and fresh taste, chile jalapeño is often used as a main ingredient and seasoner. Its pungency degree is at between 2,500 and 8,000 Scoville Heat Units, while chili green in China is under 1,000 units. If one minces chile jalapeño without a glove, the numbness it causes your fingers will last a whole day. In Mexico, lots of fresh and moderately spicy chili sauce is made from chile jalapeño.

Apart from the dishes served on the table, chili peppers are prevalent among street snacks. On Mexican streets, it's common to see vendors peddling a kind of snack you may rarely taste before. It's ice cream with chili flavor. Vendors mix a small amount of Chile jalapeño juice into ice cream, which produces a mixing taste of coolness and spiciness. The slightly sweet cream coupled with fresh chili juice not only stimulates your appetite, but also freshens your mind even if you take only one bite.

Recipe Sharing

Chicken Soup with Bamboo Fungus

Ingredients

150g chicken breast, 75g sliced tomatoes, 120g egg white, 2g MSG, 8g salt, 1,000ml consomme, 200g veiled bamboo fungus, 50g choy sum, 50g cornstarch batter, 0.5g pepper, 10mL Shaoxing cooking wine, 100mL ginger-and-scallion flavored juice, 20mL melted lard

Preparation

1. Mince the chicken breast and dislodge the tendon, add ginger-and-scallion flavored juice, egg white, salt, Shaoxing cooking wine, pepper, batter, MSG and lard, and mix them well.

2. Cut the bamboo fungus into about 6cm segments and put the mixed ingredients on the bamboo fungus segments. Put in the boiling water till they are cooked through. Transfer to the serving bowl and pour the boiling consomme over. Add the parboiled choy sum and tomato slices.

鸡蒙竹荪

食材配方

鸡脯肉 150 克、番茄片 75 克、鸡蛋清 120 克、味精 2 克、食盐 8 克、清汤 1000 毫升、竹荪 200 克、菜心 50 克、水淀粉 50 克、胡椒粉 0.5 克、料酒 10 毫升、葱姜水 100 毫升、化猪油 20 毫升

制作工艺

1. 鸡脯肉加工成泥状，去筋，加入葱姜水、鸡蛋清、食盐、料酒、胡椒粉、水淀粉、味精、化猪油搅拌成鸡糁。

2. 竹荪切成长约 6 厘米的段，将鸡糁平铺在竹荪上，入沸水锅中煮熟后捞出，盛入碗中，灌上烧沸的清汤，加入焯水后的菜心，番茄片即成。

Recipe Sharing

Cold Tomato and Cucumber Soup

Ingredients

200g cucumbers, 50g red peppers, 50g green peppers, 200g tomatoes, 20g red onions, 200g V8 mixed vegetable juice, 5g garlic, 20g fresh breadcrumbs, 20g basil leaves

Preparation

1. Wash and peel the cucumbers, red peppers, green peppers, tomatoes, red onions and garlic, cut into large pieces and blend into vegetable puree with the blender.

2. Add mix V8 vegetable juice and fresh breadcrumbs to the vegetable puree, mix well and season with salt and ground black pepper.

3. Move the soup to a refrigerator and leave to cool.

4. Top with basil leaves before transferring to a serving dish.

番茄黄瓜冻

食材配方

黄瓜 200 克、红椒 50 克、青椒 50 克、番茄 200 克、红洋葱 20 克、V8 果汁 200 克、大蒜 5 克、新鲜面包糠 20 克、罗勒叶 20 克

制作工艺

1. 将黄瓜、红椒、绿椒、番茄、红洋葱、大蒜等洗净去皮等，切成大块放入搅拌机内搅打成蔬菜泥。

2. 在蔬菜泥中加入 V8 果汁，新鲜面包糠，搅拌均匀用盐、胡椒粉调味既成。

3. 食用前将汤放入冷藏冰箱中冷藏，使温度降低。

4. 出品前用罗勒叶装饰即可。

FLAVORING

The Flavors of Chinese Cuisine

The flavors of food are of great significance to the Chinese.

Over 3,500 years ago, Yi Yin[1] translated the "philosophy of governance" into a "mouth-watering recipe". According to him, managing the country was a bit like cooking a delicious meal: the cook should produce harmony between the five flavors of sweet, sour, bitter, pungent and salty by making **meticulous** choices. Yi Yin's theory had pervaded other areas of Chinese life ever since. Later in **the Warring States Period**, a **hermit** from Chu State wrote a series of articles, in one of which he mentioned the relationship between the four flavors, namely sour, salty, sweet and bitter. He wrote, "Yin and Yang are different but they work together **harmoniously**; the tastes of sour and salty, sweet and bitter are contrary to each other, yet they are even in their part."[2] In ancient Chinese philosophy, Yin and Yang were a concept of **dualism**, describing how obviously opposite or contrary forces might actually be **complementary** and interdependent in the natural world, and how they might give rise to each other as they **interrelate** to one another. This showed how the Chinese understood the **unity of opposites** in early times. The four flavors, which later **evolved** into sour, sweet, bitter and spicy-hot, were then used as a **metaphor** for the various tastes of life as they could work with each other and create richer and more complicated tastes. Chinese people also use "Chai Mi You Yan Jiang Cu Cha" (the firewood, rice, oil, salt, soy, vinegar and tea) to refer to the **trivial** yet indispensable things in everyday life. It is obvious that for the Chinese, life is just like the taste of food — complicated, compound, **contradictory**, yet **innovative** and rich in flavors.

Traditionally, China has four schools of cuisines, namely Lu, Chuan, Cantonese and Huaiyang cuisines.

Lu cuisine, also called Shandong cuisine, features fresh and salty taste, where the spring onion is usually used to increase the fragrance of the dishes. Yinpin Tofu (Steamed Tofu Stuffed with Vegetables) and Congshao Haishen (Braised Sea Cucumbers with Spring Onions) are the most famous specialties.

Chuan cuisine, also known as Sichuan cuisine, is famous for its compound flavors. In fact, of the four schools, Sichuan cuisine has the largest number of flavors. Generally speaking, there are six basic tastes in Sichuan cuisine: sweet, sour, ma (**tingling** and numbing sensation), spicy-hot, bitter and salty, through the combination of which more than twenty main compound flavors are created. The most distinctive flavors are the fish-fragrance, spicy-hot and numbing, strange, Jiaoma and Jiaoyan flavors. The world may know Sichuan cuisine for its spicy flavor, while there are still flavors that are not spicy or numbing at all, such as litchi, sour and sweet, and fragrant sweet. These non-spicy compound flavors **constitute** a larger part of Sichuan cuisine.

Cantonese and Huaiyang cuisines share one characteristic in common: they both involve a lot of **aquatic product**, and have generally sweet and light flavors. The Western world is more familiar with Cantonese cuisine and its celebrated dishes such as Fotiaoqiang (Steamed Abalone with Shark's Fin and Fish Maw in Broth) and Baizhanji (Sliced Cold Chicken).

Huaiyang cuisine is **derived** from the native cooking styles of the region surrounding the lower reaches of the Huai and Yangtze rivers. Taking fishery products as the main ingredients, it has a combination of light, mild and sweet flavors while the original tastes of ingredients are well **preserved**. In addition, the dishes are usually well-designed and good-looking. Although it is one of the several sub-regional styles within Jiangsu cuisine, Huaiyang cuisine is widely seen in Chinese culinary circles as the most popular and **prestigious** style of Jiangsu cuisine. Two of the most celebrated dishes are Dazhu Gansi (Braised Shredded Chicken with Ham and Dried Tofu) and Songshu Guiyu (Sweet and Sour Mandarin Fish).

Notes

[1] Yi Yin: 伊尹，商朝开国元勋，杰出的政治家、思想家，中华厨祖。

[2] Yin and Yang are different but they work together harmoniously; the tastes of sour and salty, sweet and bitter are contrary to each other, yet they are even in their part. 阴阳不同气，然其为和同也；酸咸甘苦之味相反，然其为善均也。（出自《鹖冠子·环流五》）

Vocabulary

meticulous /məˈtɪkjələs/ *adj.* 极为细致的

the Warring States Period 战国时期

hermit /ˈhɜːmɪt/ *n.* 隐士

harmoniously/haːˈməʊniəsli/ *adv.* 和谐地

dualism /ˈdjuːəlɪzəm/ *n.* 二元论

complementary /ˌkɒmplɪˈmentri/ *adj.* 互补的

interrelate /ˌɪntərɪˈleɪt/ *v.* 相互关联（影响）

unity of opposites 对立统一

evolve /ɪˈvɒlv/ *v.* 演变

metaphor /ˈmetəfə(r)/ *n.* 隐喻

trivial /ˈtrɪviəl/ *adj.* 琐碎的

contradictory /ˌkɒntrəˈdɪktəri/ *adj.* 对立的

innovative /ˈɪnəveɪtɪv/ *adj.* 创新的

tingling /ˈtɪŋglɪŋ/ *adj.* 有刺痛感的

constitute /ˈkɒnstɪtjuːt/ *v.* 组成

aquatic product 水产品

derive /dɪˈraɪv/ *v.* 派生

preserve /prɪˈzɜːv/ *v.* 保持

prestigious /preˈstɪdʒəs/ *adj.* 有声望的

Tasks

 I. Group Talk

In the past, communication between different regions was difficult because back then people lacked proper means to do so. Today, we live in the era of globalization where different cultures tend to interfere and even fuse with each other. Culinary culture is no exception. Do you know anything about culinary culture that's from a foreign country and is already accepted in China? Talk with your partners and share your opinions.

 II. Writing

There must be some special dishes that could always remind you of your hometown. Do your own research, and write an article of at least 80 words to introduce a dish from your hometown, focusing on its flavoring.

_____

_____

_____

_____

_____

III. Translation

Translate the following sentences into English.

1. 山东位于黄河下游，气候温和，省内汇集有大河、大湖、丘陵、平原、大海等多样性的地貌，因此鲁菜的食材选料品种异常丰富与均衡。

_____

2. 2010年5月，四川省会成都市被联合国教科文组织授予"世界美食之都"的荣誉称号。

_____

3. 粤菜源自中原，传承了孔子所倡导的"食不厌精，脍不厌细"的中原饮食风格，因此做法比较复杂、精细。

_____

4. 淮扬菜发源于扬州、淮安，极为讲究刀工，整体口味清淡偏甜。

_____

5. 由于地理、气候、物产、文化、信仰等的差异，中餐菜肴风味差别很大，形成众多流派，有四大菜系、八大菜系之说。

_____

IV. Short Video

Pick your favorite one from the four schools of Chinese cuisine. Make a short video of no more than five minutes to introduce its famous flavors to the audience.

The Flavors of Sichuan

Passage Ⓑ

With a recorded history of more than 3,000 years and more than 1,000 rivers in its territory today, Sichuan Province has always been a significant part in Chinese history. Thanks to Li Bing's Dujiangyan Irrigation System, Chengdu, the capital city of Sichuan, has been lucky enough to maintain relatively stable development in agriculture in history. With its variety of produces, Chengdu is called the **Land of Abundance**.

During its history, Sichuan had gone through multiple mass **immigration** waves. Millions of people from nearby provinces settled down in Sichuan, bringing their local lifestyles and more importantly, cuisines. Gradually, people in different parts of the province used flavors and ingredients from their original homeland in cooking and thus had invented and developed dishes of new local **specialties**. As Sichuan people say, in Sichuan cuisine, "Each dish has its own style; a hundred dishes have a hundred different flavors." However, Sichuan cuisine is most famous for its distinctive fiery flavors. People believe that the heat brought by the spicy food could drive out the cold, wet climate from the human body.

Today, there are mainly three styles of Sichuan cuisine, namely Shanghe Bang (the upper river style), Xiahe Bang (the lower river style) and Xiaohe Bang (the small river style). Shanghe Bang is represented by Chengdu and Leshan cuisines, which have relatively light flavors and more traditional dishes. Xiahe Bang refers to Chongqing and Dazhou cuisines that are mainly spicy and innovative homemade dishes. Xiaohe Bang is featured by Zigong, Neijiang, Luzhou and Yibin cuisines, which are characterized by their strong and rich flavors.

Despite their differences, they all share one essential characteristic: the compound of flavors.

Among the many combinations, the most famous yet mysterious one may be the fish-fragrance flavor. People may understand the flavor for its literal meaning: the smell of a live fish, which is generally perceived as an unpleasant smell. But for Sichuan people, fish-fragrance flavor may not be about fish at all. It's called fish-fragrance because the seasoning for this flavor is commonly used in the cooking of fish dishes, featuring a very complicated yet balanced taste including salty, sour, sweet, spicy and a strong aroma of ginger, spring onions and garlic. This unique flavor is widely used in local dishes, such as the famous Yuxiang Rousi (Fish-Fragrance Shredded Pork). Yet the very essence of this flavor — **pickled** chili, which provides the dish with its **subtle** spicy and sour taste — is a kind of local specialty in Sichuan cuisine, making it hard for people in other parts of the country to cook **authentic** fish-fragrance flavor.

Another **featured** Sichuan flavor is Jiaoma (Sichuan pepper and spring onion) flavor, whose tingling and numbing effect are created by Sichuan pepper while its salty, fragrant and fresh tastes by spring onions. Jiaoma flavor is relatively difficult for cooks, as it requires a **swift** effort on **grinding** all the seasonings (including pepper, spring onion and salt) into paste or sauce before they lose the beautiful aroma in the air. Thus, this flavor is usually applied to cold dishes including sliced chicken and fish.

The strange flavor is also a unique member of Sichuan cuisine, where the salty, sweet, spicy-hot, numbing, sour and fragrant tastes can be found at the same time. It is called strange because it never stresses any of the tastes, but always a harmonious balance between all the flavors. One typical dish is the Strange Flavored Broad Beans.

As for the less famous non-spicy flavors, one of them is the litchi flavor, featuring a **blend** of sweet, sour and salty tastes with a **hint** of aroma from ginger, spring onion and garlic. That subtle **imitation** of litchi fruit makes the flavor a special member of Sichuan cuisine. One of the typical dishes is called Guoba Roupian (Crispy Rice with Pork Slices), in which the strong fragrance of crispy rice is softened by the sweet taste of juicy pork slices, and a hint of sour makes the flavor more adorable and **appetizing**.

There are other well-known non-spicy flavors and dishes, including the famous dish in Chinese **national banquet** — Kaishui Baicai (Steamed Chinese Cabbage in Stock), and Tianshaobai (Sweet Braised Pork with Rice) — a typical homemade dish known to every Sichuan local.

Vocabulary

Land of Abundance 天府之国

immigration /ˌɪmɪˈɡreɪʃn/ *n.* 移民

specialty /ˈspeʃəlti/ *n.* 特色菜；特产

pickle /ˈpɪkl/ *n.* 泡菜 *v.* 腌制

subtle /ˈsʌtl/ *adj.* 微妙的

authentic /ɔːˈθentɪk/ *adj.* 正宗的

featured /ˈfiːtʃəd/ *adj.* 有特色的

swift /swɪft/ *adj.* 迅速的

grind /ɡraɪnd/ *v.* 磨碎

blend /blend/ *n.* 混合物 *v.* 混合

hint /hɪnt/ *n.* 暗示；微量

crispy /ˈkrɪspi/ *adj.* 脆的

appetizing /ˈæpɪtaɪzɪŋ/ *adj.* 促进食欲的

imitation /ˌɪmɪˈteɪʃn/ *n.* 模仿

national banquet 国宴

Tasks

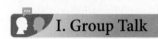 **I. Group Talk**

There are twenty-four main compound flavors in Sichuan cuisine. Find out the one you are most familiar with and share with your partners your experience with that flavor.

 II. Writing

Shanghe Bang, Xiahe Bang and Xiaohe Bang are the three main styles of Sichuan cuisine. Do your own research, pick your favorite style and write an article of at least 80 words to introduce it to the reader, focusing on its flavoring.

_____

_____

_____

_____

_____

_____

III. Translation

Translate the following sentences into English.

1. 川菜中最广为人知的就是鱼香味，它是川菜中独有的味型。

_____

_____

2. 尽管很多人认为川菜都是辣的，但实际上，不辣的菜在川菜中占了大半。

_____

_____

3. 大约 400 年前，辣椒传入中国。在此之前，中国人从花椒、生姜、茱萸中获取辣味。

_____

_____

4. 面条在四川地区极受欢迎，尤其是用酱油、辣椒调味的红汤面。

_____

_____

5. 作为小河帮菜代表之一，自贡盐帮菜以味厚、味重、味丰为其鲜明的特色，善用椒姜，料广量重，最为注重和讲究调味。

_____

_____

IV. Short Video

Some famous Sichuan dishes were brought or created with the immigration waves throughout history. Find out one dish that was from the immigrants, and make a short video of no more than five minutes to introduce it to the audience. Focus on its historical background and flavor.

Ma

—A Numbing and Tingling Sensation

Passage C

It is widely known that Sichuan cuisine is famous for its fiery flavors. However, Hunan, Yunnan, Guizhou, Jiangxi and some other provinces love spicy flavors too. Anyone familiar with Chinese cuisine could tell the differences between these cuisines: Hunan cuisine features a strong spicy flavor; Guizhou and Yunnan people prefer sour-spicy tastes; and Jiangxi cuisine focuses on a salty-spicy flavor. Compared with cuisines in these provinces, Sichuan cuisine is unique for its Mala (spicy-hot and numbing) flavor.

Mala, in culinary terms, refers to a typical compound flavor featuring spicy-hot and numbing sensations. It is a distinctive Sichuan flavor that can be found in most parts of the province, and people believe it's Ma (numbing) that defines the Sichuan cuisine.

Ma refers to a very special numbing and tingling sensation created by Sichuan pepper, sometimes called Sichuan flower pepper, named Huajiao. The best Sichuan pepper is grown in Hanyuan, Ya'an. Local Sichuan pepper bears a strong numbing flavor that could explode in your mouth, a beautiful lavender fragrance that could **enchant** your nose, a lovely aroma of **citrus** spice that could wake your appetite, and a compound of many other flavors to remind you of many other **delicate** Sichuan dishes.

But that only happens when you are used to the taste of Sichuan pepper. For people who taste Sichuan pepper for the first time, it would feel strange and uncomfortable. Put the little purplish-red pepper **husk** in your mouth and you will feel nothing in the beginning. Chew the pepper and you will feel an explosion of strong aroma in your oral and **nasal cavity**. That explosion will soon turn into a kind of tingling feeling, starting at a point in your tongue and spreading fast to the entire mouth. Swallow the chewed pepper and the tingling and numbing feeling will even go down your throat and all the way into the stomach. The flavor of Sichuan pepper is so strong that sometimes if a person eats too much of it, he would have trouble swallowing and even breathing as he couldn't feel his throat. Some local people have a **slang** for this: Ma De Geng, meaning the numbing flavor is too strong for the person to breathe smoothly. Because of these characteristics, Sichuan pepper is usually used only in small amounts—a **pinch** of ground pepper would elevate the flavor of a dish to a whole new level.

Though Sichuan pepper is used for its numbing effect today, about 400 years ago it was one of the main sources for the spicy-hot flavor. Back then, chili had not been introduced into Sichuan, but local people had the desire and passion for spicy-hot foods which they believed could drive out the body humidity caused by the **notorious** wet climate, so they used ginger, pepper and cornel to create the fiery flavor, until they had **access** to chili. Today, chili and pepper make a perfect pair for the celebrated mala flavor, and are widely applied to hundreds of well-known Sichuan dishes, such as hot pot, dried spicy beef, Maoxuewang (Duck Blood in Chili Sauce) and so on. Instead of its standard Chinese name Lajiao, Sichuan and Guizhou locals call it Haijiao (overseas pepper if translated literally), to remind people of the fact that this lovely spicy plant was introduced to these areas through **sea route**.

Note

[1] Ma De Geng：麻得哽，口语中原意指被花椒麻到吞不下东西。

Vocabulary

enchant /ɪnˈtʃɑːnt/ v. 使心醉
citrus /ˈsɪtrəs/ n. 柑橘
delicate /ˈdelɪkət/ adj. 鲜美的；精致的
husk /hʌsk/ n. 外皮
nasal /ˈneɪzl/ adj. 鼻部的
cavity /ˈkævəti/ n. 腔

slang /slæŋ/ n. 俚语
pinch /pɪntʃ/ n. 少量，一撮
notorious /nəʊˈtɔːriəs/ adj. 臭名昭著的
access /ˈækses/ n. 接触
sea route 海路

Tasks

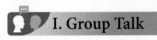 I. Group Talk

Ma is a strong taste. Share with your partners your most unforgettable experience with the numbing sensation created by Sichuan pepper, and describe it in detail.

 II. Writing

Pick a non-spicy, featured Sichuan flavor and find out its representative dishes. Write an article of at least 80 words on the flavor and the dishes, introducing their characteristics and other aspects.

_____

_____

_____

_____

_____

_____

_____

_____

_____

_____

_____

_____

_____

III. Translation

Translate the following sentences into English.

1. 麻是川菜中最为特别的味道。

_____

_____

2. 古人没有将麻味归为五味，因为他们发现这种感觉和普通味觉带给人的感受并不一样。

_____

_____

3. 现代科学认为，辣味不是一种味道，而是一种感觉。麻味也是一样。

_____

_____

4. 麻味能够带来短暂的麻痹，因此川菜中的麻辣组合也就变得有道理了：辣味刺激痛觉，而麻味使得舌头感受变迟钝，这样吃辣似乎就变得容易一些了。

_____

_____

5. 和红花椒相比，青花椒的麻更加刺鼻，适合火锅和烧菜。

_____

_____

IV. Short Video

Pick a featured Sichuan flavor and find out one representative dish that includes Ma taste. Make a short video of no more than five minutes to introduce it to the audience. Focus on its flavor and what role Ma plays in it.

Supplementary Reading

Italian Cooking

What we call flavor is in fact a composite quality, a combination of sensations from the taste buds in our mouth and the odor receptors in our nose. These sensations are chemical in nature: We taste and smell when the receptors are triggered by specific chemicals in foods. So, it's easy to understand why foods suddenly taste dull when we have a cold: The nose is blocked so the receptors cannot function properly.

From a scientific point of view, cooking is a process of both sanitizing the food and unleashing the flavoring aroma chemicals in it, so that the food becomes both delicious and easy to digest. Ancient people have long found out the key to this art: heating the food. The higher the food temperature is, the more volatile the aroma molecules become. With different cooking sequences of various ingredients and proper cooking temperature, complicated fragrances and flavors are created this way.

According to ancient people, salty, sour, sweet, bitter and pungent constitute the basic tastes of foods. However, modern science has found out that pungent spices are not actually "tasted" by taste buds, but "felt" by the irritated nerves. With a relatively mild stimulation, we could feel the numbing sensation of Sichuan pepper, Huajiao. With a much stronger stimulation, we would feel the spicy heat usually brought by chilies and peppers. Different combinations of various tastes, aromas and stimulations have created so many styles of cuisines around the world.

Like the Chinese cuisine, Italian cooking is so well-known for its variety of ingredients and rich tastes that it's one of the most popular cuisines in the world. Flavors in Italian cooking define the very base of the dishes, and they are regarded as the key architectural principle of the art of Italian cuisine. The three key techniques for an Italian cook are known as battuto (to strike), soffritto (to slightly-fry), and insaporire (to bestow taste). Battuto is also used to refer to the finely chopped mixture of uncooked ingredients (usually including butter / olive oil, parsley and onion). Then the battuto is sautéed in a skillet until it turns into a soffritto. Finally, the vegetables or other principal ingredients are added to the soffritto base and heated properly until the ingredients are completely coated with the flavor elements of the base, and the insaporire is ready.

With these three key techniques, people in different regions have created their unique styles of Italian cooking: Bolognese, Venetian, Tuscan, Sicilian and so on. Like the different schools of Chinese cuisine, these regional styles differ from each other. For example, Bolognese style features in its expensive ingredients, innovative cooking methods and constant exploration of harmonious contrast between texture and flavor. But only sixty miles away, Florentine cooks prefer dishes with unadorned themes using carefully measured ingredients. These sharp differences are mainly caused by the geographical and climatic diversity of this peninsula country.

Despite these differences, Italy's cooking is generally defined by certain ingredients. Anchovies help create heady flavors, and balsamic vinegar made from fine white grapes is put in the dish at the end of the cooking process so that its aroma will carry through into the finished dish. Other common herbs and spices such as basil, oregano, marjoram, and bay leaves are widely used in Italian cooking, along with garlic, various cheese and olive oil.

Unlike the French cuisine, there is no Italian haute cuisine, and all good Italian foods are cooked in the style of the family—the only true Italian cooking style, despite their different backgrounds. In this sense, Italian and Chinese both regard the food as an invisible thread connecting the individual to the family. Wherever a person is, the flavors of a home-style dish could always bring up the subtle nostalgia.

In the relationships of its parts, the pattern of a complete Italian meal stresses both harmony and balance: no dish overwhelms another, either in quantity or in flavor, each provides a fresh sensation of taste, color and texture, forever presenting an amazing experience to the people.

Recipe Sharing

Beef Stewed with Turnip

Ingredients

500g beef, a 10cm long turnip, some Chinese cinnamon, star anise, cinnamon leaf, pepper, ginger, 2 tablespoons of cooking wine, 2 tablespoons of light soy sauce, 1 tablespoon of dark soy sauce, some shallots and salt

Preparation

1. Rinse the beef and peeled turnip, and cut them into 5cm dices.

2. Add water in a large casserole, blanch the beef and stew it until the beef stock becomes clear.

3. Add Chinese cinnamon, star anise, cinnamon leaf, pepper, ginger, cooking wine, light soy sauce and dark soy sauce in the casserole, cover with a lid.

4. Stew with small-to-medium flame for an hour before adding the diced turnip. Stew for another 30 minutes, season with salt and chopped shallots.

白萝卜烧牛肉

食材配方

牛肉一斤、10厘米左右长的白萝卜一个、桂皮、八角、桂树叶、辣椒、生姜、料酒两勺、生抽两勺、老抽一勺、小葱三四根、盐。

制作工艺

1. 牛肉和去皮白萝卜洗净后，都切成五六厘米见方的小块。

2. 大砂锅里装满水，放入牛肉煮开，撇去血沫，直到牛肉汤变清为止。

3. 往汤中加入洗净的桂皮、八角、桂树叶、辣椒和生姜块，再倒入料酒、生抽和老抽，盖上砂锅盖。

4. 用中小火炖一个小时后，加入白萝卜块再一起炖烧半小时，根据口味浓加细盐，并撒上葱花。

Beef Stewed with Red Wine and Vegetables

Ingredients

vegetable oil, salt, black pepper ground fresh from the mill; 2 pounds of boneless beef chuck, cut into 2-inch stewing cubes; 1.5 cups of sturdy red wine, preferably a Barbera from Piedmont; 1 pound of small white onions, 4 medium carrots, 4 meaty celery stalks, 1.5 pounds of fresh peas, unshelled weight; 0.25 cup of extra virgin olive oil

Preparation

1. Put enough vegetable oil into a small sauté pan to come to 1/4 inch up the sides, and turn on the heat to medium high. When the oil is quite hot, put in the meat, in successive batches if necessary not to crowd the pan. Brown the meat to a deep color on all sides, transfer it to a plate, using a slotted spoon or spatula, put another batch of meat in the pan, and repeat the above procedure until all the meat has been well browned.

2. Pour the fat out of the pan, pour in 1/2 cup of wine, and simmer it for a few moments while using a wooden spoon to loosen the browning residues from the bottom and sides of the pan. Remove from heat.

3. Peel the onions and cut a cross into each at the root end. Peel the carrots, wash them in cold water, and cut them into sticks about 1/2 inch thick and 3 inches long. Cut the celery stalks into pieces about 3 inches long, and split them in half lengthwise, peeling away or snapping down any strings. Wash the celery in cold water. Shell the peas.

4. Choose a heavy-bottomed pot with a tight-fitting lid that can later accommodate all the ingredients of the recipe. Put in the browned meat cubes, the contents of the browning pan, the onions, olive oil, and the remaining cup of wine. Cover tightly and turn on the heat to low.

5. When the meat has cooked for 15 minutes, add the carrots, turning them over with the other ingredients. After another 90 minutes, add the celery, and give the contents of the pot a complete turn. If there is very little liquid in the pot, put in 1/2 to 2/3 cup of water. After 15 minutes, add a few

pinches of salt, liberal grindings of pepper, and turn over the contents of the pot. Continue cooking until the meat feels tender. Altogether, the stew should take about 2 hours to cook. Taste and correct for salt and pepper before serving.

Ahead-of-time Note:

Like all stews, this one will have excellent flavor when prepared a day or two in advance. Reheat gently just before serving.

红酒蔬菜炖牛肉

食材配方

植物油、盐、现磨黑胡椒；2 磅去骨牛颈肉，切成 5 厘米见方的块状；1.5 杯烈性红葡萄酒，以意大利皮埃蒙特产的巴贝拉酒（Barbera）为佳；1 磅白色小洋葱、4 个中型胡萝卜、4 根粗芹菜茎、1.5 磅未剥的新鲜豌豆、1/4 杯特级初榨橄榄油

制作工艺

1. 取一小煎锅，加入足量植物油至 1/4 高度，开到中高火。油热后，加入牛颈肉块翻炒至所有面均为深色后，用勺子或铲子放入盘中。如煎锅放不下，可将牛肉分批翻炒至所有牛肉全部上色。

2. 倒出锅中油脂，锅中加入 1/2 杯红酒，文火加热同时用木勺将锅内壁的褐变残留物铲松。之后将锅放置一旁备用。

3. 洋葱去皮，在每一个洋葱根部切十字形。胡萝卜去皮，冷水洗净后切成约 1.2 厘米厚、7.5 厘米长的条状。芹菜茎切成约 7.5 厘米长的段后从中间掰开去筋，冷水洗净。豌豆剥好备用。

4. 取一厚底大锅（带密封较好的锅盖），足够装下所有食材为宜。往锅内放入煎好的牛肉块，煎锅内的液体，洋葱，橄榄油，以及剩下的红酒。紧紧盖上锅盖，小火加热。

5. 肉煮至 15 分钟时，加入胡萝卜，将锅内食材翻搅后继续加热。再等 90 分钟后，加入芹菜，将锅内食材全部翻搅均匀。如锅内液体很少，可加入 1/2 至 2/3 杯清水。15 分钟后，加入少许盐和现磨黑胡椒粉，搅匀。继续加热至肉质变软。整体上，制作这道菜一共需要大约两小时。出菜前加入盐和胡椒粉调味。

注意： 同其他炖菜类似，这道菜应提前 1 ~ 2 天制作，以获取最佳风味。上菜前再次加热即可。

COOKING METHODS

The History of Chinese Cooking Methods

Passage A

About 80% of the current Chinese dishes are cooked by "stir-frying". Therefore, when talking about Chinese cooking methods, many people may take it for granted that stir-frying must be the oldest one invented by the Chinese people. In fact, many other cooking methods had been invented long before the invention of "stir-frying" in Chinese history.

In ancient times in China, people did not find the use of "fire", so the meat eaten at that time was basically raw. For example, the Chinese character "Kuai" usually referred to raw sliced or shredded fish.

Then people began to use fire and eat roasted food. The method of roasting was actually very popular during early Tang Dynasty in China. At that time, people would put a goose stuffed with glutinous rice and seasonings into the belly of a whole sheep, sew up the belly of the sheep, and then roast it as a whole. At that time, roasting was called "Zhi" in Chinese and "Kuai" and "Zhi" were both popular cooking methods. If you could eat raw fish slices with roasted meat at that time, it was really a pleasure in life. Therefore, this should be the origin of the Chinese idiom—Raw fish slices and roasted meat **suit everyone's taste**.

Another cooking method which was invented at the same period of time with roasting should be "boiling". Chinese people have been using boiling to cook food for a long time. "Ding", an ancient Chinese cooking vessel with two loop handles and three or four legs was specially used for boiling food. After that, many more cooking methods were derived from "boiling". For example, "**stewing**", the most typical cooking method in the north and northeastern China, is derived from "boiling". When a family has a reunion dinner, Dongbei Stewing Food would be a best choice for people in the north. It's really a rare happy moment in life for them.

Simmering, the favorite cooking method for people in Fujian and Guangdong, is also derived from "boiling". A large bowl of simmering soup can not only strengthen the body, **dispel** dampness and **replenish** "Qi" (vitality), but also bring a strong feeling of hometown and family.

Later, Chinese ancients invented a new cooking method — **steaming**. It is said that "steaming" had been invented as early as in the era of the Yellow Emperor. Archaeologists found the earliest steamer (4,600 years from now) "Zeng" in the Shijiahe site in Tianmen, Hubei Province, which coincides with the era of the Yellow Emperor. With such a long history and profound cultural heritage, Hubei's steamed dishes now are renowned all over the world.

When talking about deep-frying, it was invented later than roasting, boiling and steaming. After all, oil was not easily obtained in ancient times. Not until the period of Three Kingdoms did deep-fried food began to appear on the Chinese table. At the Northern and Southern Dynasties, deep-frying method had been developed quite maturely.

At the time of the Northern and Southern Dynasties, after thousands of years of development of Chinese civilization, "stir-frying", the future major cooking method in China, was finally on the stage. Moreover, compared with roasting, boiling and steaming, stir-frying is more time-saving and fuel-saving. As a result, it was rising rapidly and has been strongly loved by the general public in China up to now.

Vocabulary

suit everyone's taste 脍炙人口

stewing /'stjuːɪŋ/ *n.* 炖

dispel /dɪ'spel/ *v.* 驱散

replenish /rɪ'plenɪʃ/ *v.* 补充

steaming /'stiːmɪŋ/ *n.* 蒸

Tasks

 I. Group Talk

China is a big country of broad and profound catering culture. For a long time, due to the influence of geographical environment, climate, produces, cultural traditions and national customs, some regions have formed unique styles of cooking with certain local flavors. Among them, Guangdong cuisine, Sichuan cuisine, Shandong cuisine, Jiangsu cuisine, Zhejiang cuisine, Fujian cuisine, Hunan cuisine and Anhui cuisine are known as the "eight famous cuisines". The cooking methods of China's eight famous cuisines have their unique charms. Discuss with your partners about different cooking methods of the eight famous cuisines and give examples of famous and representative dishes made by these cooking methods.

 II. Writing

In China, cooking is a kind of art. Cooking in China, like music, dance, poetry and painting, has the great significance of improving the realm of life. There are various Chinese cooking methods: slicing, stewing, steaming, frying, grilling, wire drawing, etc. But it is quite different in the Western countries. Learn more about the differences between Chinese and Western cooking methods and write an article of about 80 words.

_____

_____

_____

_____

_____

_____

_____

III. Translation

Translate the following sentences into English.

1. 中国"八大菜系"的烹调技艺各具风韵，其菜肴之特色也各有千秋，主要的烹饪方法就有十四种之多。

   _____

   _____

2. 中国烹饪是世界上最早利用热能的领域。因此，蒸法也便被世界上称为中国蒸。

   _____

   _____

3. 几千年来，受到各地环境、气候、物产、风俗以及饮食习惯的影响，中国菜形成了多种多样的烹饪方法。

   _____

   _____

4. 人学会用火将食物烤熟标志着人类烹饪的开端。最初，原始人类将食物直接放在火上炙烤，这种烹饪方式被后人称为"火烹法"。

   _____

   _____

5. 在夏商周时期，随着青铜器的发明和大量使用，加热炊具和锋锐的切割刀具的发明，以及厨师调味技术的进步，中国烹饪的技术三大要素由此产生。

   _____

   _____

IV. Short Video

Make a short video to introduce the cooking method of one of your favorite Chinese dishes, and share it with your classmates.

The Origin of the Unique Chinese Cooking Method

—Stir-Frying

Passage B

When it comes to the cooking methods of food, most people may be able to say one or two. For example, steaming, boiling, stewing, frying and roasting, which have been widely used in various countries. However, there is also a unique Chinese cooking method, that is stir-frying. It can be said that stir-frying is the essence of Chinese cooking method. About 80% of Chinese cuisines are stir-fried, so Westerners generally believe that "stir-frying makes Chinese cuisines richer".

However, stir-frying had not been invented until thousands of years of development of Chinese civilization and the invention of other cooking methods. When it was just created during the period of Northern and Southern Dynasties, stir-frying was not popular. Even so, **scrambled eggs**, a well-known classic stir-fried food, was recorded in *Qimin Yaoshu*[1], one of the earliest **monographs** on agriculture in the world and the most complete one existing in China.

It was in the Song Dynasty that stir-frying became popular on a large scale. At that time, the iron smelting technology had reached a fairly high level, and the manufacturing technology of the **iron wok** was becoming more and more sophisticated. The popularity of wok with uniform heating, fast conduction and easy operation by cooks has laid a foundation for stir-frying.

Moreover, delicious and cheap vegetable oil has also promoted the development of stir-frying. As we all know, stir-frying needs much more cooking oil than boiling or stewing. In ancient times, oil mainly came from animals, and animal oil was **scarce** and expensive, which ordinary people could not afford. In the Eastern Han Dynasty, vegetable oil was invented. However, the types of vegetable oil were very limited, mainly including **almond oil** and sesame oil, and the supply was also very scarce. Ordinary people had no access to it. Later in the Northern and Southern Dynasties, especially in the Tang Dynasty, thanks to the frequent exchanges with foreign countries, the types of vegetable oil began to increase while the price gradually declined.

Zhuang Chuo[2], a well-informed man of the Northern Song Dynasty, recorded in his book *Jilei*[3] that people in the Song Dynasty used more kinds of oil in their daily life, and that they had an unprecedented enthusiasm for fried food. Shen Kuo[4] also mentioned in his *Dream Stream Essays*[5] that people in Song Dynasty loved deep-fried food so much that they even wanted to fry sesame in the sesame oil. It can be sure that the supply of vegetable oil in Song Dynasty was abundant and the price should not be high. This provided an important condition for the rapid rise of stir-frying in Song Dynasty.

In addition, it is said that before the Tang Dynasty, due to the small population, people preferred meat than vegetables. Since the Song Dynasty, China's population began to surge, but the source of meat gradually decreased. In order to solve the problem of meat shortage, vegetables began to be popular. Then people found that stir-frying could stimulate the most delicious taste of meat combined with vegetables.

Compared with steaming, boiling and roasting, stir-frying is more time-saving and fuel-saving. As a result, stir-frying rose rapidly and soon became the most popular cooking method in China since then. Moreover, it can be said that the invention of stir-frying is a major event in the history of world cooking. If there is no stir-frying, there will be no chance for the whole world to enjoy delicious Chinese food.

Notes

[1] *Qimin Yaoshu*：《齐民要术》大约成书于北魏末年（533 年—544 年），是北朝北魏时期，南朝宋至梁时期，中国杰出农学家贾思勰所著的一部综合性农学著作，也是世界农学史上的专著之一，是中国现存最早的一部完整的农书。

[2] Zhuang Chuo: 庄绰（约 1079 年—？），字季裕，惠安县人。北宋年间考证学家、民俗学家、天文学家、医药学家，尤其对灸法有深入研究。

[3] *Jilei*: 庄绰所著《鸡肋编》三卷，后人推为与《齐东野语》相埒。

[4] Shen Kuo: 沈括（1031 年—1095 年），字存中，号梦溪丈人，汉族，杭州钱塘县（今浙江杭州）人，北宋官员、科学家。

5 *Dream Stream Essays*：北宋科学家、政治家沈括（1031 年—1095 年）撰写的《梦溪笔谈》，是一部涉及古代中国自然科学、工艺技术及社会历史现象的综合性笔记体著作。该书在国际上亦受重视，英国科学史家李约瑟将其评价为"中国科学史上的里程碑"。

Vocabulary

scrambled eggs 炒鸡蛋

monograph /ˈmɒnəɡrɑːf/ *n.* 专著

iron wok 铁锅

scarce /skeəs/ *adj.* 缺乏的

almond oil 杏仁油

Tasks

 I. Group Talk

Stir-fried dishes account for a large proportion in Chinese cooking. Some people think that stir-fried dishes can be used as the basic feature to distinguish Chinese dishes from foreign dishes. After a long period of development, stir-frying has been divided into many types. For example, "plain-frying" means stir-frying without other ingredients, such as Plain-Fried Broccoli. Can you think of other types of stir-frying and stir-fried dishes? Talk with your partners and think of more examples.

 II. Writing

Stir-frying is ordinary but very important in Chinese cooking. So why do Chinese like stir-fried dishes? Learn more about stir-frying, try to analyze the reasons and write an article of about 80 words.

_____

_____

_____

_____

_____

_____

_____

III. Translation

Translate the following sentences into English.

1. 魏晋南北朝之际，炒用于做菜的明确的文字记载已经出现。这是中国乃至于世界菜肴史、烹饪史上的大事。

   _____

   _____

2. 炒菜是中国菜的常用制作方法，是中国家庭日常最广泛使用的一种烹饪方法，是将一种或几种菜在锅中炒熟的过程。

   _____

   _____

3. 在炒菜发明以前，我国古代吃东西以烧烤和煮食为主，这种原始的烹饪方式一直到唐朝还是主流，因此，古人就费了很大心思研究更加美味的烧烤做法。

   _____

   _____

4. 在唐朝之前，最早的炒菜出现在北朝时期著名农学家贾思勰的《齐民要术》上，而且是一道很简单的菜——炒鸡蛋，这算是中国炒菜的老祖宗了。

   _____

   _____

5. 明清时期，花生油的提炼方法也从北美传入我国，因为花生油非常适合炒菜，使炒菜又得到了进一步发展。

   _____

   _____

IV. Short Video

Make a short video to introduce a famous stir-fried dish in China, and share it with your classmates and friends.

The Innovation of Sichuan Cuisine Cooking Methods

Passage C

With the rapid development of China's catering industry, Chinese and Western foods are constantly **integrated**. Meanwhile, chefs are trying to promote the continuous development of Sichuan cuisine with the spirit of innovation.

For example, after mastering the essence of the baking method, Sichuan cuisine chefs make some innovative changes. Instead of using Western sauces, they use various Sichuan sauces to **marinate** the ingredients before baking, which not only maintains the taste of the dishes, but also enriches the flavor.

Moreover, in recent years, with a large number of seafood raw materials entering Sichuan, the cooking method of **blanching** has also been accepted by most people in Sichuan. Combined with the characteristics of Sichuan cuisine, chefs have developed a series of new blanching dishes, such as Blanching Flowering Cabbage and Blanching Whelk Meat with Sichuan-Flavored Dipping Sauce.

Meanwhile, with the increasingly frequent exchanges between China and foreign countries, and the rapid improvement of industrialization in today's society, the rhythm of people's life and work is accelerating. People pay more attention to nutrition, health and quality of food. Modern Sichuan cuisine enterprises have realized this. They have made innovations in cooking tools and methods, and have found a way to combine Chinese and Western cuisines. Various restaurants in Chengdu are actively exploring reasonable and efficient cooking methods and operation procedures instead of sticking to the tradition process. They learn from the phased and integrated cooking methods of Western food, and apply it to Sichuan cuisine.

For example, Chen Wei, the owner of a famous Sichuan restaurant in Chengdu, has always been committed to integrating Sichuan cuisine with the Western cooking methods he has learned. He called this integrated cuisine "a new style of Sichuan cuisine", that is, learning from the cooking methods of Western cuisine and presenting the Sichuan dishes to diners in a more beautiful, scientific and nutritious way. He believes that Sichuan cuisine often uses the cooking methods of rapid stir-frying in the environment of high temperature and hot oil, which made the nutrients get lost with the **volatilization** of water. Therefore, he learned from the Western low-temperature slow cooking methods, trying his best to maintain the original taste and nutrients of the ingredients, and finally presented a more delicate food taste of the Sichuan cuisine. A dish in his restaurant, called "Beef & Sichuan Pickles", is to process beef with the commonly used **roasting** method of Western food, accompanied by Sichuan spicy sauce and famous Sichuan pickles. This dish, which looks like a Western cuisine in appearance, hides the traditional Sichuan flavor inside. At present, this innovative cooking method has been widely accepted and won the general favor of customers in Chengdu.

Furthermore, there are more and more communications between Chinese and Western cooking methods. For example, you can eat authentic salads and beef steaks in a Sichuan restaurant, and you can also eat a variety of delicacies with both Chinese and Western elements. To sum up, the cooking methods of Sichuan cuisine are constantly innovating and improving. Through this continuous progress, Sichuan cuisine has kept up with the development of the times and moved towards modernization. It is also the modernization and innovation that will make Sichuan cuisine move from the southwest to the whole country and even the whole world, and to a more brilliant future.

Vocabulary

integrate /ˈɪntɪɡreɪt/ v. 综合

marinate /ˈmærɪneɪt/ v. 腌制

blanching /ˈblɑːntʃɪŋ/ n. 白灼

volatilization /ˌvɒlətɪlɪˈzeɪʃn/ n. 蒸发

roasting /ˈrəʊstɪŋ/ n. 烤

Tasks

 I. Group Talk

At present, many Sichuan restaurants are trying to create a new style of Sichuan cuisine with unique flavors. For example, when you open the menu of Daronghe, a famous Sichuan restaurant in Chengdu, you can see not only traditional Sichuan dishes such as Mapo Tofu, Double-Cooked Meat, Fish-Flavored Shredded Pork and so on, but also new dishes with innovative integration. Take Daronghe's signature dish "Kaimenhong" (A Good Start) as an example, which is actually learning from "Fish Head with Chopped Bell Pepper" in the classic Hunan cuisine, using chili pepper and Sichuan pepper instead of chopped bell pepper. Discuss with your partners about new styles of Sichuan cuisine and give more examples of your favorite new Sichuan dishes.

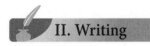 II. Writing

At present, in addition to Sichuan cuisine, other Chinese cuisines are also actively innovating and integrating with Western cuisines to meet the tastes of modern consumers and keep pace with the international standards. Some people believe that Chinese cuisine should maintain its own tradition, while others believe that it should pursue innovation. What's your point of view? Write an article of about 80 words.

_____

_____

_____

_____

_____

_____

_____

III. Translation

Translate the following sentences into English.

1. 过去的一百多年在其他领域没有创新的时候，中国的餐饮却诞生了如此之多的创新，这其实是值得我们深入思考的。

   _____

2. 为了让羊肉膻味变淡，四川人的做法是提前把羊肉煮熟，用羊骨头熬出的一锅味道浓郁的高汤来烫，再在锅底里加葱、姜和陈皮，吃起来反倒是比北方的羊肉火锅还要清淡些。

   _____

3. 川菜小吃原本是街边摊，难登大雅之堂。经过餐厅的整理、包装拔高，做得更雅致，更适合白领消费，算是以形式取胜。

   _____

4. 中国的饮食其实是一个大量创新的过程，与其他行业相比，中国人把更多的聪明才智都用在了吃上面。

   _____

5. 川菜所谓的百菜百味，包括家常味、麻辣味、香辣、鱼香等，这些口味都是以辣椒为根基来进行的创新，辣椒带来了产品创新和技术创新。

   _____

IV. Short Video

Make a short video to introduce one of your favorite new Sichuan dishes, and share it with your classmates and friends.

Supplementary Reading

Differences between Chinese and Western Cooking Methods

Western cooking methods are not as complicated and changeable as China's. The whole cooking process in the Western countries is strictly in accordance with scientific norms, and the methods are standardized. Therefore, the work of the chef has become extremely monotonous mechanical work. Although it seems monotonous, the simple process enables them to achieve amazing success in large-scale commercialization. Hamburgers and pizza can even keep the same taste in global chain stores. Although the preparation of these foods is much simpler than that of Chinese food, Westerners still make great efforts to achieve the goal of uniform taste.

Take McDonald's fries as an example. In order to obtain the best state of French fries, at the beginning, the experimenters focused on studying the humidity, cooking and storage time of French fries. They found that the best state of French fries was closely related to the storage time of potatoes in the store and the oil temperature. Later they found that the perfect storage time for potatoes was three weeks. After that, they found that the processing method had a greater impact on the quality of French fries. Therefore, the engineers conducted numerous tests in the laboratory through a year's research, and finally found that the oil temperature would drop sharply when the cold French fries were poured into a hot pot at 325 degrees Fahrenheit. But no matter how much the oil temperature dropped, the best state of the French fries was when the oil temperature rose back again to 3 degrees Fahrenheit higher than that of the French fires. According to this principle, they designed an automatic French fries machine to ensure the uniform taste of the French fries.

In contrast, cooking is an art in China. It is interesting and even has a certain playfulness, which deeply attracts the Chinese people who take food as the supreme joy of life. Just like music, dance, poetry and painting, cooking in China has the great significance of improving the realm of life. There are numerous cooking methods in Chinese cuisine: stewing, steaming, frying, grilling, braising, boiling and so on. The dishes made by these cooking methods are dazzling and tasty.

Unlike Western food, cooking methods of Chinese cuisine are changeable. We can say that it has developed to a higher stage with a kind of flexibility. Each dish can be developed and adjusted on the basis of the original one to adapt to different regions, seasons, objects and functions. For example, fried beef with mushrooms should be darker in color and heavier in taste in winter, and lighter in color and taste in summer; For people in Jiangsu Province, sugar can be added to the seasoning, but for customers in Sichuan and Hunan, more chillies should be added. Therefore, without the flexibility and variety of cooking methods, the unique charm of Chinese cooking will be lost.

Recipe Sharing

Translucent Beef Slices

Ingredients

500g beef, 1,500g cooking oil for deep-frying, 5g salt, 5g sugar, 5g ground roasted Sichuan pepper, 10g ground chilies, 2g five-spice powder, 2g MSG, 5g sesame oil

Preparation

1. Cut the beef into slices, sprinkle with salt and lay out to dry.

2. Scatter the dry beef slices on a grill, and then toast over charcoal fire to further evaporate the water content. Transfer the toasted slices to a steamer and steam for about 60 minutes. Cut the beef slices into thin slices of 6cm long and 4cm wide while they are still hot.

3. Heat oil in a wok to 120 ℃ and deep-fry the beef slices. Remove, and add the ground chilies, ground roasted Sichuan pepper, sugar, MSG and five-spice powder. Blend well. Wait till the beef cools, add sesame oil and mix well.

灯影牛肉

食材配方

黄牛肉 500 克、食用油 1500 克（约耗 80 克）、食盐 5 克、白糖 5 克、花椒粉 5 克、辣椒粉 10 克、五香粉 2 克、味精 2 克、芝麻油 5 克

制作工艺

1. 牛肉滚片成厚薄均匀的大片，撒上食盐，晾干水分。

2. 把晾好后的牛肉平铺在架上，放进炉内用炭火烘干，再上笼蒸约 60 分钟，趁热切成长约 6 厘米、宽约 4 厘米的片。

3. 锅内下食用油烧至 120℃，放入牛肉片炸透捞出，加辣椒粉、花椒粉、白糖、味精、五香粉翻簸均匀，晾凉后加芝麻油拌匀即成。

Recipe Sharing

Caramelized Banana Bread Pudding with Rum Anglaise

Ingredients

200g croissant, 250mL cream, 2 eggs, 40g sugar, 60g pineapples, 100g bananas, 10g raspberries, 40mL rum

Preparation

1. Caramelized bananas: cut bananas into pieces, heat sugar and butter in a pan till they melt into caramel, add banana pieces and fry till coated with caramel syrup, add a little rum and mix well to make caramelized bananas. Slice the bottom of the pineapple, sprinkle with icing sugar and bake in a 200℃ oven for 5 minutes.

2. Cut the croissant into small pieces, sprinkle with icing sugar and bake in a 200 ℃ oven for 5 minutes.

3. Heat the milk, sugar and vanilla in a saucepan over medium heat until boiling. Strain and leave to cool to room temperature. Add an egg, an egg yolk and sugar into a stainless steel bowl, whisk till white, pour in the warm milk, stir well and strain.

4. Spread the softened butter inside the pudding mold, sprinkle with sugar and spread well, put the croissant pieces from Step 2 and banana pieces from Step 1 into the mold, pour in the milk from Step 3, compact and cover with the pineapple bottom.

5. Fill the baking tray with hot water, place the pudding moulds and bake at 180℃ for 20 minutes.

6. Rum anglaise: Put the vanilla sticks into the milk and bring to a boil, remove from the heat and set aside. Add the egg yolks and sugar into a stainless steel bowl and whisk till white, pour in the hot milk and stir well, heat again and stir till boiling, turn down the heat till thicken, add rum and remove from the heat to make the rum anglaise.

7. When serving the dish, demould the pudding, serve to the plate and drizzle the rum anglise.

焦糖香蕉面包布丁

食材配方

牛角包 200 克、奶油 250 毫升、鸡蛋 2 个、白砂糖 40 克、菠萝 60 克、香蕉 100 克、覆盆子 10 克、朗姆酒 40 毫升

制作工艺

1. 制作焦糖香蕉：将香蕉切块，锅中加糖和黄油加热融化成焦糖，放入香蕉块煎制，至香蕉裹匀焦糖浆，加少许朗姆酒拌匀成焦糖香蕉备用。将菠萝底部切片，撒糖粉后，入 200℃烤炉烤 5 分钟备用。

2. 将牛角面包切成小块，撒糖粉后，入 200℃烤炉烤 5 分钟备用。

3. 锅中倒入牛奶、白糖和香草条，中火加热至沸腾，过滤后静置至室温，另取不锈钢盆，加全蛋 1 个、蛋黄 1 个、白糖搅打至发白，倒入凉温的牛奶拌匀后过滤。

4. 布丁模具内涂抹化软的黄油，撒入白糖铺匀，将 2 的面包丁和 1 的香蕉块放入模具内，倒入 3 的牛奶，压实后最后加菠萝底封面。

5. 烤盘内注入热水，放入布丁模具，180℃烤炉烤 20 分钟备用。

6. 制作朗姆酒英式奶油汁：将牛奶加香草条煮沸，离火备用。另取不锈钢盆，加蛋黄、白糖搅打至发白，倒入热牛奶搅匀，再次上火加热，边加热边搅动，至煮沸后转小火煮稠，加入朗姆酒调味，离火成朗姆酒英式奶油汁。

7. 上菜时，布丁脱模装盘，淋汁即成。

CHAPTER IX NOTABLE FOODS

Notable Foods in China's Eight Famous Cuisines

Passage A

Chinese food, which boasts a long history, has been rising from the most basic demand of human beings to culinary culture.

During the Republican Period, the food culture in China has developed greatly and the "eight famous cuisines" in China had formed. The cooking skills of the "eight famous cuisines" have their own charms, and the characteristics of their dishes are also different. There are thousands of notable foods in China's typical local dishes. They have **exquisite** materials and different flavors. They pay attention to the **coordination** and unity of color, aroma, taste, shape and utensils, and enjoy a high reputation in the world.

For hundreds of years, countless notable foods have spread from the civil society to the palace, and then spread throughout the country and the world. Such as the Beijing notable foods "Beijing Roast Duck", "**Kirin Tofu**", Tianjin "Goubuli Steamed Buns", Hangzhou "Dongpo Meat", "West Lake Vinegar Fish", Shanghai "Songjiang Perch", Hunan "Dong'an Chicken", "**Steamed Preserved Meat**", Hubei "**Calipash and**

Calipee Soup with White Gourd", Anhui "Wuwei Smoked Duck", and Sichuan "Mapo Tofu" and "**Diced Chicken with Red Pepper**", are all famous delicacies in the world.

A variety of these notable foods not only reflect exquisite traditional skills, but also contain various beautiful and moving legends or **allusions**, which have become an important part of China's food culture. Take Goubuli Steamed Buns in the city of Tanjin as an example. These popular buns are all of the same size and handmade. When served in neat rows on a tray, they look like budding chrysanthemum flowers. The wrapping is thin, the fillings juicy, the meat tender and the taste delicious and not at all greasy. There is an interesting story behind it. Goubuli Steamed Buns are first sold in Tianjin about 150 years ago. A local man by the name of Gouzi (dog) worked as an apprentice in a shop selling Baozi (Steamed Buns). After three years, he set up his own shop. Because his buns were so delicious, he soon had a thriving business with more and more people coming to buy his buns. Soon, the demand exceeded supply. As a result, people had to wait for a long time to buy his buns. Impatient, some people would call out to urge him on, but as he was so busy preparing his buns, he didn't answer. People therefore call his buns Goubuli, which means Gouzi pays no attention. This eccentric name, however, has had very good promotional effects, and has been used ever since. Goubuli is now a time-cherished brand name in Tianjin.

Moreover, in Zhejiang cuisine, there is a well-known dish called Dongpo Meat. This dish was named after Su Dongpo, a great poet of the Northern Song Dynasty, who created it when he was an official in Hangzhou. It is said that, when he was in charge of the drainage work for the West Lake, Su Dongpo rewarded workers with stewed pork in soy sauce, and people later named it Dongpo Meat, to commemorate this gifted and generous poet.

Furthermore, the notable foods in China are not only a food culture, but also a geographical culture. The formation and development of a cuisine is the result of the comprehensive effects of specific regional products, climate, living customs, etc. Although the old saying that "governing a big country is like cooking a small dish" is talking about the way of ruling a state, it also shows that the formation and development of a cuisine is as complex and profound as governing a country, and its rise and fall are closely related to the social, political and economic development.

Vocabulary

exquisite /ɪkˈskwɪzɪt/ *adj.* 精致的

coordination /kəʊˌɔːdɪˈneɪʃn/ *n.* 协调

Kirin Tofu 麒麟豆腐

Steamed Preserved Meat 腊味合蒸

Calipash and Calipee Soup with White Gourd 冬瓜鳖裙

Diced Chicken with Red Pepper 辣子鸡丁

allusion /əˈluːʒn/ *n.* 典故

Tasks

 ### I. Group Talk

Most of Chinese notable foods, which have been inherited by famous chefs for several generations, have their own development history. They not only reflect exquisite traditional skills, but also have various beautiful and moving legends or allusions, which have become an important part of China's food culture. Discuss with your partners about the legends or allusions you know behind the notable foods of the eight famous cuisines.

 ### II. Writing

Among various notable foods of the eight famous cuisines, which one is your favorite and what are the reasons? Write an article of about 80 words.

_____

_____

_____

_____

_____

_____

_____

III. Translation

Translate the following sentences into English.

1. 早在春秋战国时期，中国传统饮食文化中南北菜肴风味就表现出了差异。到唐宋时，南食、北食各自形成体系。

   _____

   _____

2. 一个菜系的形成和它的悠久历史与独到的烹饪特色是分不开的，同时也受到这个地区的自然地理、气候条件、资源特产、饮食习惯等影响。

   _____

   _____

3. 中国人饮食习俗的一大特点是就是使用筷子。筷子在中国有着悠久的历史，在古代，它被称为"箸"。此外，东方各国使用的筷子大多也都源自中国。

   _____

   _____

4. 经过千百年的实践与锤炼，中国菜系非常讲究各种技巧，从厨师的刀工火候，到盛菜的花色式样，花样百出，绝无仅有。

   _____

   _____

5. 天南地北，口味各异，东甜西酸，南辣北咸是一个大致的区分，但互相之间多有融合，界限并不那么明显。

   _____

   _____

   _____

IV. Short Video

Make a short video to introduce the stories behind a notable food in China, and share it with your classmates.

Chinese Noodles

In the world's food culture, noodles can be seen almost everywhere. In Italy, each person consumes about 30 kg of noodles every year. The popular Italian food, "**macaroni**", was introduced by Marco Polo, a famous Italian **merchant** and adventurer, when he returned to Italy from his trip to China, and it has become an Italian specialty. Japanese noodles are also passed down from China, also known as "ramen" in Japan, a transliteration from Chinese. Indonesian people call noodles "Mi", which is borrowed from Minnan language "noodles". Suodo Noodles, a famous food of Indonesia, is actually beef (or chicken) noodle soup mixed with Indonesia's unique sweet soy sauce, white vinegar and chili sauce. It tastes sweet, sour and spicy, which is quite delicious.

Noodles were originally known as "Tangbing" (noodles in the soup). The records of Tangbing can be found in the historical data of Han Dynasty, but this does not mean that Chinese people began to eat noodles from Han Dynasty. There was a time when the origin of noodles was debated all over the world. The Italians thought noodles originated in Italy, and the Arabs insisted noodles come from Arabia, while the Chinese said noodles actually were invented in China. Later, Chinese **archaeologists** discovered the oldest noodles in the Lajia site in Qinghai Province. The noodles were golden in color and slender in shape. They were very similar to today's ramen. This discovery shows that China had noodles more than 4,000 years ago, thus finally ended the debate on the origin of noodles in the world.

Although noodles are delicious, they are not easy to store and carry. Later, through **trial and error**, Chinese people finally solved this problem and invented a kind of noodles which was easy to store and carry, that is, dried noodles. The earliest record of dried noodles in history is in the classic novel **Water Margin**[1]. In Chapter 54, among the gifts given to Yang Xiong, a character in the *Water Margin*, by the **licentious** monk Pei Ruhai, there are "some noodles and a few bags of Beijing dates". *Water Margin* was written in the late Yuan Dynasty and the early Ming Dynasty. It can be seen that dried noodles has a history of at least 600 years.

Chinese people have the custom of eating noodles on their birthday. The noodles eaten on birthdays are called birthday noodles, which have the meaning of good health and a long life. If people offer birthday congratulations to an elder man, they usually choose "Yigenmian" (**single-strand** noodle) to wish him a long life like noodles. In addition to birthday noodles, some places also have the custom of eating noodles during the Spring Festival and on the second day of February in Chinese lunar calendar. In addition, noodles often appear as a lucky food when people are newly married or on a celebration dinner party of a baby's completion of first month of life.

Noodles of every place in China boast their own unique characteristics. For example, Shanxi Daoxiao Noodles, Lanzhou Noodles, Old Beijing Noodles with Soy Bean Paste, Henan's Stewed Noodles, Shanghai's Yangchun Noodles, Sichuan's Dandan Noodles, Shandong's Yifu Noodles, Wuhan Hot-Dry Noodles with Sesame Paste and so on are all popular noodles among Chinese people. It's hard to say which of them is better, but each has its own **merits**.

Note

[1] *Water Margin*：《水浒传》是元末明初施耐庵编著的章回体长篇小说。《水浒传》是中国古典四大名著之一，问世后，在社会上产生了巨大的影响，成了后世中国小说创作的典范。《水浒传》是中国历史上最早用白话文写成的章回小说之一，流传极广，脍炙人口；同时也是汉语言文学中具备史诗特征的作品之一，对中国乃至东亚的叙事文学都有深远的影响。

Vocabulary

macaroni /ˌmækəˈrəʊni/ *n.* 通心粉

merchant /ˈmɜːtʃənt/ *n.* 商人

archaeologist /ˌɑːkiˈɒlədʒɪst/ *n.* 考古学家

trial and error 反复试验

licentious /laɪˈsenʃəs/ *adj.* 放肆的

Water Margin 《水浒传》

single-strand /ˈsɪŋglstrænd/ *adj.* 单根的

merit /ˈmerɪt/ *n.* 优点

Tasks

 I. Group Talk

Noodles have a long history in China. As mentioned in the passage, Chinese people would eat noodles on many special occasions, such as on one's birthday. Moreover, besides these traditions, there are many folk stories behind noodles. Discuss with your partners about the story you know about noodles.

 II. Writing

There are many kinds of noodles in China, such as the Old Beijing Noodles with Soy Bean Paste, Wuhan Hot-Dry Noodles with Sesame Paste, Lanzhou Noodles, Shanxi Daoxiao Noodles, Chengdu Dandan Noodles, etc. Which one is your favorite and why? Write an article of about 80 words.

_____

_____

_____

_____

_____

_____

_____

_____

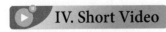

III. Translation

Translate the following sentences into English.

1. 面条是一种制作简单，食用方便，营养丰富，既可作为主食又可作为快餐的健康食品，早已为世界人民所接受与喜爱。

2. 面条作为中国北方居民的主食，从餐厅到一般平民百姓，制作面条方法及吃法的花样之繁多，堪称世界一流。

3. 南方人一般将面条作为早晚点或两餐之间的加餐，其作用相当于点心。北方地区则不同，面条一般作正餐主食，并常年食用，久吃不厌。

4. 中华民族创造的面条之所以能传至意大利和日本、朝鲜，扩展至欧洲和阿拉伯世界，是因为我们民族的经济文化影响力曾经很大，我们民族曾为很多国家的人所仰视。

5. 方便面绝对是日常大家吃得最多的一种方便美食了，它快速简单，几分钟即可出锅，完全不需要你有多么精湛的厨艺，只需要把料包放进去煮即可便能享受到美味的面食。

IV. Short Video

Make a short video to introduce the stories you know behind noodles in China, and share it with your classmates.

Sichuan Hot Pot

If you **wander** around Chengdu on a summer's evening, you'll find the streets filled with diners, their tables being scattered on the pavement in the shade of leafy trees. Many will be gathered around bubbling hot pots, in which **multitudes** of dried chilies and Sichuan pepper bob up and down in an oily red broth, and they'll happily while away the hours chatting, sipping beer and dipping morsels of food into the broth to cook. Hot pot — literally "fire-pot" — is one of Sichuan's most popular dishes, and a favorite excuse for a get-together with family or friends. Fancy restaurants **lavish** with **marble** and **chandeliers** serve it up for the rich, while beneath the steel supports of city flyovers, jobbing laborers crouch to eat it, paying a few Jiao a time for a skewer of food to plunge into the spicy broth.

The "hot pot" itself, a wok or saucepan filled with a rich, oily broth **resplendent** with chiles, sits on a simple stove, whether it's a charcoal brazier on the pavement, a gas burner in a specially cut-out restaurant table, or just an electric ring on the floor at home. Around it, like stars around the moon, are arranged dozens of small plates, each being piled high with a different ingredient. Esoteric cuts of offal, many varieties of mushroom, bamboo shoots, ribbons of celtuce, crisp green leaves and sweet potato noodles are usually favored, but high-class restaurants will also serve exotic seafood and whole little fish. Guests select pieces of food from this tempting array and plunge them into the soup to cook. Some ingredients need only be briefly scalded, and others are simmered for several minutes. When the tidbits are ready, they are fished out with chopsticks and dipped into sesame oil seasoned with garlic, salt and MSG, and sometimes also chopped scallions and cilantro.

The Sichuanese love to spend whole days or evenings eating hot pot, sitting around the simmering wok for hours on end. The end of the meal is comfortably undefined: you just cook and eat, and eat and cook, for as long as you fancy. The pace ebbs and flows, with bursts of enthusiastic guzzling followed by gentle lulls of inactivity. Even when the meal is naturally drawing to a close, there is usually someone still exploring the broth for a forgotten tidbit.

If you watch a group of Sichuanese people eating hot pot, you'll notice that they tend to spurn the lean meat and shrimp balls favored by foreigners, pouncing instead on the **slithery** or rubbery offal delicacies: pimpled gray sheets of tripe, jagged strips of ox or pig aorta, curling duck intestines, rabbit kidneys and coils of pig intestine.

Hot pot makes a fiery winter dish, potent at expelling the creeping dampness of the Sichuan winter. But the Sichuanese are peculiar among the Chinese in eating their spicy hot pot all year round. Even at the height of summer, hot pot restaurants will be overflowing with customers fanning themselves in the sweltering heat even as they swallow another mouthful of chile-laden food. The casual informality and **delirious** heat of hot pot encourage raucous behavior; popular hot pot restaurants are themselves like cauldrons full of **cacophony** and laughter. Eating Sichuan's numbing-and-hot hot pot produces the most delicious physical sensation, a warmth and relaxation that begins in the belly and radiates out to all the extremities of the body, soothing away tension and anxiety, calming the mind and spirits.

Vocabulary

wander /ˈwɒndə(r)/ *v.* 漫步

multitude /ˈmʌltɪtjuːd/ *n.* 大量

lavish /ˈlævɪʃ/ *adj.* 奢华的

marble /ˈmɑːbl/ *n.* 大理石

chandelier /ˌʃændəˈlɪə(r)/ *n.* 烛台

resplendent /rɪˈsplendənt/ *adj.* 光亮的

slithery /ˈslɪðəri/ *adj.* 滑溜的

delirious /dɪˈlɪriəs/ *adj.* 特别愉快的

cacophony /kəˈkɒfəni/ *n.* 刺耳的声音

Tasks

 I. Group Talk

Hot pot is the representative of Chinese food and Sichuan hot pot is also the epitome of Sichuan food culture. It originated in Sichuan, and red, the main color of hot pot, symbolizes enthusiasm, bravery and hope. Hot pot is extremely popular among people in Sichuan, China and even the whole world. Discuss with your partners about the reasons why Sichuan hot pot is welcomed by people from all parts of China and the whole world.

 II. Writing

If one of your foreign friends is going to visit Chengdu. He wrote a letter to ask you to recommend the popular food here. Write a letter in reply of about 80 words to recommend the best food or restaurants in Chengdu and state your reasons.

_____

_____

_____

_____

_____

_____

_____

_____

_____

_____

III. Translation

Translate the following sentences into English.

1. 当下，成都火锅火遍海内外，成为大西南这个"世界美食之都"的又一张光彩名片。

   _____

   _____

2. 在四川，上到官员，下到百姓，无一不偏爱四川火锅，家家都会做。

   _____

   _____

3. 重庆人吃火锅的豪放与气吞山河之势是其他地区无法相比的，这正是巴渝饮食文化的体现，是古老巴民族勇武豪放性格和饮食文化心理的表现。

   _____

   _____

4. 火锅之乐，在于意趣，亲朋好友，宾客同伴，围着火锅，边煮边烫，边吃边聊，可丰可俭，其乐无穷。

   _____

   _____

5. 火锅的消费与中餐相比低得多。作为请客或家庭就餐也不失面子，而且火锅不受季节影响，春夏秋冬皆可以吃，味道也五花八门。

   _____

   _____

IV. Short Video

Make a short video to introduce the stories you know behind Sichuan hot pot, and share it with your classmates.

Supplementary Reading

Turkish Cuisine

Turkish cuisine, along with French cuisine, Chinese cuisine and Italian cuisine, is one of the top cuisines in the world. Turkey's natural advantages in geographical location bring up rich and colorful food with many special tastes. Over the past few centuries, Turkish cuisine has been evolving, mainly because it has undergone many cultural and political changes, such as the Byzantine Empire period and Ottoman Empire period.

The diverse geographical conditions of different regions in Turkey contribute to the complexity of the country's cuisine. For example, the eastern part of the Black Sea is not suitable for wheat production due to heavy rainfall. Therefore, residents have developed dishes that mainly rely on corn and corn flour. Similarly, the southeastern part of Anatolia is famous for delicious kabobs because of its abundant livestock. Moreover, the Aegean Sea region is well known for producing olives, and thus its olive oil vegetable dishes and medicinal materials are popular.

Istanbul has always been attracting a large number of immigrants from other parts of Turkey, who moved to the city in order to find jobs. Because of this, Istanbul has become the cultural center of Turkey, sharing the most delicious Turkish food from all regions of this country.

Turks in the Ottoman Empire used to eat two meals a day. The first meal was eaten between morning and noon, more like brunch. The second meal will be eaten at any time between the afternoon and evening. This meal usually consists of meat dishes with vegetables and beans. However, most families now enjoy three meals a day. The breakfast on weekdays is simple and fast, but on weekends, the whole family will get together in the morning to enjoy a big meal. Turkey's lunch usually consists of seasonal dishes, soup, salad, etc. Meat-based dishes and desserts that require time and effort to prepare are not common in this meal. In contrast, dinner is usually more exquisite and abundant, because only at this time, family members who have worked for a whole day will get together again.

In Turkey, there is also an unofficial meal called "yatsilik", which is eaten around 9 or 10 o'clock after dinner. Nuts, dried fruits and fresh fruits are usually served with Turkish black tea. Some of the most common foods in "yatsilik" are seasonal fresh fruits, dried plums, figs, dried fruit pulp (grape, apricot or mulberry), and nuts such as pistachios, almonds, roasted chickpeas, roasted pumpkin seeds and sunflower seeds, walnuts and hazelnuts.

Furthermore, the chef will wish diners at the table "afiyet olsun" before eating in Turkey, which is basically equivalent to saying "wish you a good appetite". In return, diners will also say "elinize saglik", which literally means "wish your hands healthy". But it is more accurate to interpret it as a praise for the chef, which means "very delicious, well done".

Recipe Sharing

Steamed Pork with Sweet Stuffing

Ingredients

500g pork belly with skin attached, 100g glutinous rice, 100gred bean paste, 1,000g cooking oil for deep-frying, 50g brown sugar, 20g lard, 50g sugar, 5g caramel color

Preparation

1. Clean the pork, boil in water till cooked through. Remove the pork, smear its skin with caramel color while it is still hot, and then leave to drain. Deep-fry the pork in the cooking oil till its skin bubbles and browns.

2. Boil the glutinous rice in water till soft, remove and then mix well with brown sugar, caramel color and lard.

3. Cut the pork into slices about 8cm long, 3cm wide and 0.4cm thick. Make one cut in each slice to sandwich the red bean paste. Lay the sandwiched slices into a steaming bowl, making sure that the skin is in

contact with the base of the bowl. Cover the pork slices with glutinous rice, steam for two hours and remove. Turn the bowl upside down to transfer its contents onto a serving dish. Sprinkle with the sugar.

甜烧白

食材配方

带皮猪五花肉 500 克、糯米 100 克、豆沙 100 克、食用油 1000 克（约耗 20 克）、红糖 50 克、猪油 20 克、白糖 50 克、糖色 5 克

制作工艺

1. 猪五花肉刮洗干净，入汤锅中煮熟后捞出，趁热在皮上抹一层糖色，晾干水分，放入油锅中炸至表皮起泡、上色后待用。

2. 糯米淘洗干净，放入沸水锅中煮至无硬心时捞出，趁热加入红糖、糖色、猪油拌匀。

3. 将上色后的猪肉切成长约 8 厘米、宽约 3 厘米、厚约 0.4 厘米的夹层片，每片肉中夹上豆沙，皮向下装入蒸碗，上面填入糯米，上笼蒸 2 小时，食用时翻扣装入盘内，撒上白糖即成。

Recipe Sharing

Velvet Chicken Soup with Mushrooms and Scallion Oil

Ingredients

Main Ingredients: 80g red onions, 80g scallions, 150g shiitake mushrooms, 100g enoki mushrooms, 250g white mushrooms, 50g celeries, 50g butter, 1 litre chicken stock, 400mL light cream, 50g butter, 50g flour, some salt and ground pepper

Chicken Stock (1 litre): 1kg chicken bones, 2 kg water, 100g carrots, 50g onions, 50g leeks, 30g celeries, 10g dried shiitake mushrooms, 10g garlic, 1 bunch of spices (bay leaf, thyme), 1 clove

Auxiliary Ingredients: 60g mushrooms, 30g celeries, 30g scallions, 30g raw chicken breasts, 45mL olive oil, 30g green onions, some salt and ground pepper

Preparation

1. Chicken stock: Boil the chicken bones, add water and bring to a boil, skim off the foam, add seasoned vegetables such as carrots, onions, leeks and celery, dried mushrooms, garlic, spice bunches and cloves, cook for 2 hours over low heat to make chicken stock, filter and cool in ice water and set aside.

2. Fry the diced raw chicken breast, mushrooms, celery and scallion in olive oil till aromatic, season with salt and ground pepper, keep warm and set aside when ready, serve in a seasoned spoon; make scallion oil: blanch the scallion, cool and season with olive oil, salt and pepper and then ground with a blender to make scallion oil and set aside.

3. Mushroom velouté chicken soup: Fry red onion and scallion with butter, add shiitake mushroom, enoki mushroom, white mushroom and celery till aromatic, add white flour and fry to make white roux, add chicken stock and boil to thicken, smash with a blender, pour into a pot and add light cream and bring to a boil, season and make mushroom velouté chicken soup, keep warm and set aside.

4. Presentation: Pour the hot mushroom soup into the soup pot, drizzle in the scallion oil to enhance the aroma, add the mushroom ingredients into the seasoning spoon and serve.

葱油蘑菇天鹅绒鸡汤

食材配方

主料：红葱 80 克、大葱 80 克、香菇 150 克、金针菇 100 克、白口蘑 250 克、西芹 50 克、黄油 50 克、鸡高汤 1 升、淡奶油 400 毫升、黄油 50 克、白面粉 50 克、盐和胡椒粉适量

鸡高汤（1 升）：鸡骨架 1 千克、水 2 千克、胡萝卜 100 克、洋葱 50 克、韭葱 50 克、芹菜 30 克、干香菇 10 克、大蒜 10 克、香料束（香叶、百里香）1 束、丁香 1 个

配料： 蘑菇 60 克、芹菜 30 克、大葱 30 克、生的鸡胸 30 克、橄榄油 45 毫升、青葱 30 克、盐和胡椒粉适量

制作工艺

1. 制作鸡高汤： 将鸡骨架汆水后，加清水煮沸，撇去浮沫，加胡萝卜、洋葱、韭葱、西芹等调味蔬菜，干香菇、大蒜、香料束、丁香等，用小火煮 2 小时成鸡高汤，过滤后用冰水冷却后备用。

2. 配料制作： 将切成小粒的生鸡胸肉、蘑菇、芹菜、大葱用橄榄油炒香后，加盐和胡椒粉调味，保温备用（成菜时盛入调味勺）；制作葱油：大葱焯下水，冷却后与橄榄油、盐、胡椒调和后用破壁机打碎成葱油备用。

3. 蘑菇天鹅绒鸡汤制作： 将红葱和大葱用黄油炒香，加入香菇、金针菇、白口菇、西芹炒匀，加入白面粉炒香不变色，成白色油面酱，加入鸡汤煮稠，用破壁机搅碎后，倒入锅中加入淡奶油煮沸，调味后成蘑菇天鹅绒鸡汤，保温备用。

4. 装盘： 将热的蘑菇汤倒入汤盅内，淋入葱油增香，炒香的蘑菇配料盛入调味勺中，上菜即成。

GLOSSARY

abacus /ˈæbəkəs/ n. 算盘 U1

abstract /ˈæbstrækt/ adj. 抽象的 U5

access /ˈækses/ n. 接触 U7

acclaim /əˈkleɪm/ v. 称赞 U1

accumulation /əˌkjuːmjəˈleɪʃn/ n. 积累 U6

acne /ˈækni/ n. 痤疮；粉刺 U5

acrobatics /ˌækrəˈbætɪks/ n. 杂技 U3

affective /əˈfektɪv/ adj. 情感的；表达感情的 U2

allusion /əˈluːʒn/ n. 典故 U9

almond oil 杏仁油 U8

anaerobic fermentation 厌气发酵；无氧发酵 U4

analogous /əˈnæləgəs/ adj. 相似的 U1

ancestor /ˈænsestə(r)/ n. 祖先 U2

anemic /əˈniːmɪk/ adj. 贫血的 U5

appetite /ˈæpɪtaɪt/ n. 胃口 U1

appetizing /ˈæpɪtaɪzɪŋ/ adj. 促进食欲的 U7

apprentice /əˈprentɪs/ n. 学徒 U4

aquatic product 水产品 U7

archaeologist /ˌɑːkiˈɒlədʒɪst/ n. 考古学家 U9

aroma /əˈrəʊmə/ n. 香味 U1

arthritis /ɑːˈθraɪtɪs/ n. 关节炎 U6

astronomy /əˈstrɒnəmi/ n. 天文学 U2

attach /əˈtætʃ/ v. 连接；附上 U5

auspicious /ɔːˈspɪʃəs/ adj. 吉祥的 U3

authentic /ɔːˈθentɪk/ adj. 正宗的 U7

backbone /ˈbækbəʊn/ n. 支柱 U6

bang /bæŋ/ v. 敲击 U3

banquet /ˈbæŋkwɪt/ n. 宴会 v. 宴请 U2

biorhythm /ˈbaɪəʊrɪðəm/ n. 生物周期；生物节律 U2

blade /bleɪd/ n. 刀刃；刀片 U4

blanching /ˈblɑːntʃɪŋ/ n. 白灼 U8

blend /blend/ n. 混合物 v. 混合 U7

bold /bəʊld/ adj. 大胆的 U1

bride /braɪd/ n. 新娘 U2

bridegroom /ˈbraɪdgruːm/ n. 新郎 U2

bronzeware /ˈbrɒnzweə(r)/ n. 青铜器 U5

broth /brɒθ/ n. 浓汤 U1

buckle /ˈbʌkl/ v. 扣住 U4

bun /bʌn/ n. 小圆面包 U1

cacophony /kəˈkɒfəni/ n. 刺耳的声音 U9

calamity /kəˈlæməti/ n. 灾祸 U3

Calipash and Calipee Soup with White Gourd 冬瓜鳖裙 U9

carbon steel 碳钢 U4

casserole /ˈkæsərəʊl/ n. 砂锅 U4

categorize /ˈkætəgəraɪz/ v. 分类 U1

cavity /ˈkævəti/ n. 腔 U7

celebration /ˌselɪˈbreɪʃn/ n. 庆祝 U2

censer /ˈsensə(r)/ n. 香炉 U3

ceramic /səˈræmɪk/ n. 陶 U3

chandelier /ˌʃændəˈlɪə(r)/ n. 烛台 U9

China National Image Global Survey 中国国家形象调查 U1

circulation /ˌsɜːkjəˈleɪʃn/ n. 循环；流动 U2

citrus /ˈsɪtrəs/ n. 柑橘 U7

clay /kleɪ/ n. 黏土；陶土 U4

cleaver /ˈkliːvə(r)/ n. 菜刀 U4

clink /klɪŋk/ v. 碰杯 U3

commence /kəˈmens/ v. 开始 U3

commitment /kəˈmɪtmənt/ n. 信奉；承
诺 U2

communal /kəˈmjuːnl/ adj. 公用的 U3

complementary /ˌkɒmplɪˈmentri/ adj. 互补
的 U7

complexity /kəmˈpleksəti/ n. 复杂性 U1

comprise /kəmˈpraɪz/ v. 包括；由……构
成 U5

condensed milk 炼乳 U1

condolence /kənˈdəʊləns/ n. 哀悼；慰
问 U2

constituent /kənˈstɪtʃuənt/ adj. 构成的 U3

constitute /ˈkɒnstɪtjuːt/ v. 组成 U7

contractive /kənˈtræktɪv/ adj. 收缩的；有
收缩性的 U5

contradictory /ˌkɒntrəˈdɪktəri/ adj. 对立
的 U7

conviction /kənˈvɪkʃn/ n. 确信；信仰 U3

cookware /ˈkʊkweə(r)/ n. 炊具；厨房用
具 U4

coordination /kəʊˌɔːrdɪˈneɪʃn/ n. 协调 U9

copious /ˈkəʊpiəs/ adj. 大量的 U3

cornel /ˈkɔːnəl/ n. 山茱萸 U6

counteract /ˌkaʊntərˈækt/ v. 抵消；中
和 U2

counterbalance /ˌkaʊntəˈbæləns/ v. 抵
消 U3

crisp /krɪsp/ adj. 脆的 U4

crispy /ˈkrɪspi/ adj. 脆的 U7

crisscross /ˈkrɪskrɒs/ v. 纵横交错 U5

culinary /ˈkʌlɪnəri/ adj. 烹饪的 U6

custom /ˈkʌstəm/ n. 习惯；惯例；习
俗 U2

dampness /ˈdæmpnəs/ n. 湿气 U6

debone /diːˈbəʊn/ v. 将……去骨 U4

deep-frying /ˈdiːpfraɪɪŋ/ n. 炸 U4

delicate /ˈdelɪkət/ adj. 美味的 U1 / 鲜美
的；精致的 U7

delirious /dɪˈlɪriəs/ adj. 特别愉快的 U9

demonstrate /ˈdemənstreɪt/ v. 显示 U6

derive /dɪˈraɪv/ v. 派生 U7

descendant /dɪˈsendənt/ n. 后代 U3

dice /daɪs/ v. 切丁 U4

Diced Chicken with Red Pepper 辣子鸡
丁 U9

dietary culture 饮食文化 U2

discontent /ˌdɪskənˈtent/ n. 不满 U3

dispel /dɪˈspel/ v. 驱散 U8

distinct /dɪˈstɪŋkt/ adj. 不同的 U1

distinctive /dɪˈstɪŋktɪv/ adj. 独特的 U6

diverse /daɪˈvɜːs/ adj. 多样的 U1

diversify /daɪˈvɜːsɪfaɪ/ v.（使）多样化 U6

drain /dreɪn/ v. 使排出 U5

dress up 盛装打扮 U2

dry-roasting /ˈdraɪrəʊstɪŋ/ n. 干烧 U4

dualism /ˈdjuːəlɪzəm/ n. 二元论 U7

embrace /ɪmˈbreɪs/ v. 拥抱；欣然接受 U2

enchant /ɪnˈtʃɑːnt/ v. 使心醉 U7

endorphin /enˈdɔːfɪn/ n. 内啡肽 U6

enjoy a square meal 饱餐一顿 U2

enthusiasm /ɪnˈθjuːziæzəm/ n. 热情 U1

entice /ɪnˈtaɪs/ v. 诱惑 U5

epitomize /ɪˈpɪtəmaɪz/ v. 体现 U3

essence /ˈesns/ n. 本质；实质 U2 / 精华
U6

etiquette /ˈetɪket/ n. 礼仪 U3

even /ˈiːvn/ adj. 偶数的；均衡的 U2

evolve /ɪˈvɒlv/ v. 演变 U7

excrete /ɪkˈskriːt/ v. 排出 U1

expansion /ɪkˈspænʃn/ n. 扩大；增加 U5

explorer /ɪkˈsplɔːrə(r)/ *n.* 探险家 **U6**

exquisite /ɪkˈskwɪzɪt/ *adj.* 精致的 **U9**

extrude /ɪkˈstruːd/ *v.* 挤压 **U1**

fade out 渐渐淡出；逐渐淡出 **U4**

family reunion 家庭团聚 **U2**

featured /ˈfiːtʃəd/ *adj.* 有特色的 **U7**

ferment /fəˈment/ *v.* 发酵 **U6**

fertile /ˈfɜːtaɪl/ *adj.* 肥沃的；富饶的 **U5**

festivity /feˈstɪvəti/ *n.* 欢庆；庆典 **U2**

fiddlehead fern 像小提琴头的蕨类植
物 **U5**

filial piety *n.* 孝顺 **U3**

filling /ˈfɪlɪŋ/ *n.* 馅；填充 **U1**

flat /flæt/ *n.* 刀面 **U4**

flavor /ˈfleɪvə/ *n.* 风味；口味 **U2**

fluffy /ˈflʌfi/ *adj.* 松软的 **U1**

folk wisdom 民间智慧 **U2**

folklore /ˈfəʊklɔː(r)/ *n.* 民间风俗 **U3**

formulation /ˌfɔːmjuˈleɪʃn/ *n.* 形成 **U2**

freeze off 冻坏 **U2**

frightful /ˈfraɪtfl/ *adj.* 可怕的 **U3**

from scratch 从无到有；从头做起；白
手起家 **U4**

fungus /ˈfʌŋɡəs/ *n.* 菌类（复数为
fungi） **U5**

galore /ɡəˈlɔː(r)/ *adj.* 大量的 **U3**

gamut /ˈɡæmət/ *n.* 全范围；全过程 **U3**

garlic chive 韭菜 **U1**

gastrointestinal /ˌɡæstrəʊɪnˈtestɪnl/ *adj.* 肠
胃的 **U1**

gastronomic /ˌɡæstrəˈnɑːmɪk/ *adj.* 美食
的 **U5**

gastronomy /ɡæˈstrɒnəmi/ *n.* 美食 **U1**

gourd /ɡʊəd/ *n.* 葫芦类属植物 **U5**

gourmet /ˈɡʊəmeɪ/ *n.* 美食家 **U1**

green onion 葱 **U1**

grind /ɡraɪnd/ *v.* 磨碎 **U7**

grocery /ˈɡrəʊsəri/ *n.* 食品杂货店 **U5**

handover /ˈhændəʊvə(r)/ *n.* 交接；移
交 **U2**

harmoniously /hɑːˈməʊniəsli/ *adv.* 和谐
地 **U7**

harness /ˈhɑːnɪs/ *v.* 控制（利用）（自然
力等） **U5**

heat-resistant /ˈhiːt rɪzɪstənt/ *adj.* 耐热的 **U4**

hermit /ˈhɜːmɪt/ *n.* 隐士 **U7**

hierarchical /ˌhaɪəˈrɑːkɪkl/ *adj.* 等级制度
的 **U3**

high-pitched voice 高音；调门高的嗓
音 **U4**

hint /hɪnt/ *n.* 暗示；微量 **U7**

homophonous /həˈmɒfənəs/ *adj.*（词语）
同音异形的；同音异义的 **U3**

hospitality /ˌhɒspɪˈtæləti/ *n.* 好客 **U3**

humid /ˈhjuːmɪd/ *adj.* 潮湿的 **U6**

humiliating /hjuːˈmɪlieɪtɪŋ/ *adj.* 丢脸的 **U3**

husk /hʌsk/ *n.* 外皮 **U7**

hygiene /ˈhaɪdʒiːn/ *n.* 卫生 **U3**

imitation /ˌɪmɪˈteɪʃn/ *n.* 模仿 **U7**

immerse /ɪˈmɜːs/ *v.* 沉浸 **U1**

immigration /ˌɪmɪˈɡreɪʃn/ *n.* 移民 **U7**

in lieu of 替代 **U1**

in return 作为回报 **U2**

in turn 相应地；转而 **U2**

incense /ˈɪnsens/ *n.* 香 **U3**

inclusive /ɪnˈkluːsɪv/ *adj.* 范围广泛的 **U1**

incorporate /ɪnˈkɔːpəreɪt/ *v.* 合并 **U1**

indicate /ˈɪndɪkeɪt/ *v.* 预示；表明 **U2**

indispensable /ˌɪndɪˈspensəbl/ *adj.* 不可或
缺的 **U6**

infuse /ɪnˈfjuːz/ *v.* 注入 **U1** /（辣椒面、茶
叶等）被浸渍；被泡 **U5**

ingredient /ɪnˈɡriːdiənt/ *n.* 食材 U1

inheritance /ɪnˈherɪtəns/ *n.* 继承 U3

innovative /ˈɪnəveɪtɪv/ *adj.* 创新的 U7

intangible /ɪnˈtændʒəbl/ *adj.* 无形的；非物质的 U1

integrate /ˈɪntɪɡreɪt/ *v.* 使完整；成为一体 U2 / 综合 U8

interrelate /ˌɪntərɪˈleɪt/ *v.* 相互关联（影响）U7

intestinal /ɪnˈtestɪnl/ *adj.* 肠的 U6

iron wok 铁锅 U8

irrational /ɪˈræʃənl/ *adj.* 不合理的 U3

irrigation /ˌɪrɪˈɡeɪʃn/ *n.* 灌溉 U5

jar /dʒɑː(r)/ *n.* 坛子 U4

Kirin Tofu 麒麟豆腐 U9

knife sharpener 磨刀匠 U4

knife skill 刀工 U4

ladle /ˈleɪdl/ *n.* 瓢子 U4

Land of Abundance 天府之国 U7

landlocked /ˈlændlɒkt/ *adj.* 内陆的 U6

larder /ˈlɑːdə(r)/ *n.*（家中的）食品贮藏室 U5

lavish /ˈlævɪʃ/ *adj.* 奢华的 U9

lazy Susan（餐桌上盛食物便于取食的）圆转盘 U3

leavening agent 发酵剂 U1

legend /ˈledʒənd/ *n.* 传奇 U6

lethargic /ləˈθɑːdʒɪk/ *adj.* 无精打采的；懒洋洋的 U5

lethargy /ˈleθədʒi/ *n.* 无精打采 U6

liberal /ˈlɪbərəl/ *adj.* 慷慨的 U1

licentious /laɪˈsenʃəs/ *adj.* 放肆的 U9

longevity /lɒnˈdʒevəti/ *n.* 长寿；寿命 U2

macaroni /ˌmækəˈrəʊni/ *n.* 通心粉 U9

macrobiotic /ˌmækrəʊbaɪˈɒtɪk/ *adj.* 养生饮食的；延年益寿的 U5

main course 主菜 U2

mainstay /ˈmeɪnsteɪ/ *n.* 支柱 U1

make a fortune 发财；赚大钱 U2

make toasts to 祝酒；敬酒 U2

manipulate /məˈnɪpjuleɪt/ *v.* 操作 U1

marble /ˈmɑːbl/ *n.* 大理石 U9

marinate /ˈmærɪneɪt/ *v.* 腌制 U8

martial arts *n.* 武术 U3

Maya /ˈmeɪə/ *n.* 玛雅人 U6

merchant /ˈmɜːtʃənt/ *n.* 商人 U9

merge /mɜːdʒ/ *v.* 融入 U1

merit /ˈmerɪt/ *n.* 优点 U9

metabolism /məˈtæbəlɪzəm/ *n.* 新陈代谢 U5

metaphor /ˈmetəfə(r)/ *n.* 隐喻 U7

metaphorical /ˌmetəˈfɒrɪkl/ *adj.* 比喻的 U3

meticulous /məˈtɪkjələs/ *adj.* 极为细致的 U7

microcosm /ˈmaɪkrəʊkɒzəm/ *n.* 缩影 U3

microorganism /ˌmaɪkrəʊˈɔːɡənɪzəm/ *n.* 微生物 U4

mill /mɪl/ *v.* 磨成粉 U1

mince /mɪns/ *v.* 切碎 U4

mixture /ˈmɪkstʃə(r)/ *n.* 混合物 U6

moisten /ˈmɔɪsn/ *v.* 使湿润 U5

moldy /ˈməʊldi/ *adj.* 发霉的 U6

monograph /ˈmɒnəɡrɑːf/ *n.* 专著 U8

multitude /ˈmʌltɪtjuːd/ *n.* 大量 U9

mustard /ˈmʌstəd/ *n.* 芥菜 U5

mysterious /mɪˈstɪəriəs/ *n.* 神秘的 U6

napa cabbage 大白菜 U1

nasal /ˈneɪzl/ *adj.* 鼻部的 U7

national banquet 国宴 U7

navigator /ˈnævɪɡeɪtə(r)/ *n.* 航海家 U6

niche /niːʃ/ *adj.* 小众的 U3

norm /nɔːm/ *n.* 常态；标准；准则 U5

notorious /nəʊˈtɔːriəs/ *adj.* 臭名昭著
的 U7

observance /əbˈzɜːvəns/ *n.* 遵守 U3

occasion /əˈkeɪʒn/ *n.* 重大场合 U2

omelette /ˈɒmlət/ *n.* 蛋饼；煎蛋卷 U2

oomph /ʊmf/ *n.* 活力 U5

originate /əˈrɪdʒɪneɪt/ *v.* 发源 U6

osmanthus /ɒzˈmænθəs/ *n.* 桂花；木犀属
植物 U2

palate /ˈpælət/ *n.* 味觉 U1

paramount /ˈpærəmaʊnt/ *adj.* 至关重要
的 U1

partake of 吃；喝 U3

paste /peɪst/ *n.* 膏；肉（或鱼等）酱 U4

patriarchal /ˌpeɪtriˈɑːkl/ *adj.* 家长制的 U3

peddle /ˈpedl/ *v.* 沿街叫卖 U4

peristalsis /ˌperɪˈstælsɪs/ *n.* 蠕动 U6

personality /ˌpɜːsəˈnæləti/ *n.* 性格；个
性 U2

phenomenon /fɪˈnɒmɪnən/ *n.* 现象（复数
为 **phenomena**）U2

phlegm /flem/ *n.* 痰；黏液 U5

pickle /ˈpɪkl/ *n.* 泡菜 *v.* 腌制 U7

pictorial /pɪkˈtɔːriəl/ *adj.* 图画的；照片
的 U5

pimple /ˈpɪmpl/ *n.* （尤指长在脸上的）粉
刺；小脓包 U5

pinch /pɪntʃ/ *n.* 少量；一撮 U7

plague /pleɪg/ *v.* 困扰；折磨 U5

plump /plʌmp/ *adj.* 丰满的；胖乎乎的 U5

pockmark /ˈpɒkmɑːk/ *n.* 麻子 U3

pomelo /ˈpɒmələʊ/ *n.* 柚子 U5

porcelain /ˈpɔːsəlɪn/ *n.* 瓷 U3

portend /pɔːˈtend/ *v.* 预示 U3

potsticker /ˈpɒtstɪkə/ *n.* 锅贴 U1

pottery /ˈpɒtəri/ *n.* 陶器 U4

poultry /ˈpəʊltri/ *n.* 家禽 U5

practice /ˈpræktɪs/ *n.* 做法；活动 U2

predominantly /prɪˈdɒmɪnəntli/ *adv.* 占主
导地位地；显著地 U5

prefecture /ˈpriːfektʃə(r)/ *n.* 县；辖区；地
方官 U5

presage /ˈpresɪdʒ/ *n.* 预兆 U3

preserve /prɪˈzɜːv/ *v.* 保持 U7

prestigious /preˈstɪdʒəs/ *adj.* 有声望的 U7

prevent /prɪˈvent/ *v.* 妨碍；阻止 U2

prickling /ˈprɪklɪŋ/ *adj.* 刺痛的 U5

produce /ˈprɒdjuːs/ *n.* 物产 U1

property /ˈprɒpəti/ *n.* 属性 U3

proprieties /prəˈpraɪətiz/ *n.* 礼节；行为规
范 U3

prospect /ˈprɒspekt/ *n.* 前景；前途 U1

prosperity /prɒˈsperəti/ *n.* 繁荣 U2

protect... against... 保护……使免
受…… U2

protocol /ˈprəʊtəkɒl/ *n.* 礼节 U3

pun /pʌn/ *n.* 双关语 U3

pungency /ˈpʌndʒənsi/ *n.* 辛辣 U1

quench /kwentʃ/ *v.* 淬火 U4

quick-witted /ˌkwɪkˈwɪtɪd/ *adj.* 机敏的 U2

radical /ˈrædɪkl/ *n.* （汉字）偏旁；部首 U5

reddish /ˈredɪʃ/ *adj.* 微红的 U6

refreshing /rɪˈfreʃɪŋ/ *adj.* 清爽的 U4

regions south of the Yangtze River 长
江以南地区 U2

relieve /rɪˈliːv/ *v.* 解除；减轻 U2

reminiscent /ˌremɪˈnɪsnt/ *adj.* 引人联想
的 U3

render /ˈrendə(r)/ *v.* 使成为；使变得；使
处于 U5

replenish /rɪˈplenɪʃ/ *v.* 再度装满 U2 /
补充 U8

symbolize /ˈsɪmbəlaɪz/ *v.* 象征 **U6**

terrain /təˈreɪn/ *n.* 地形；地势 **U5**

texture /ˈtekstʃə(r)/ *n.* 质地；口感 **U1**

the Warring States Period 战国时期 **U7**

therapeutic /ˌθerəˈpjuːtɪk/ *adj.* 使人镇静的；使人放松的 **U5**

therapy /ˈθerəpi/ *n.* 治疗；疗法 **U5**

thrive /θraɪv/ *v.* 兴旺；欣欣向荣 **U5**

tingling /ˈtɪŋlɪŋ/ *adj.* 有刺痛感的 **U7**

tissue /ˈtɪʃuː/ *n.* 组织 **U6**

toon /tuːn/ *n.* 椿树 **U1**

tot /tɒt/ *v.* 合计 **U1**

traditional Chinese medicine 传统中药 **U2**

translate /trænsˈleɪt/ *v.* 转化 **U6**

trial and error 反复试验 **U9**

trivial /ˈtrɪviəl/ *adj.* 琐碎的 **U7**

tuber /ˈtjuːbə(r)/ *n.* （马铃薯等植物的）块茎 **U5**

turnip /ˈtɜːnɪp/ *n.* 白萝卜；芜菁 **U5**

undignified /ʌnˈdɪɡnɪfaɪd/ *adj.* 无尊严的 **U3**

undoubtedly /ʌnˈdaʊtɪdli/ *adv.* 毫无疑问地 **U2**

unity of opposites 对立统一 **U7**

utensil /juːˈtensl/ *n.* 用具 **U1** / 器皿 **U4**

utterance /ˈʌtərəns/ *n.* 说出；表达 **U3**

vacuity /vəˈkjuːəti/ *n.* 虚；贫乏 **U5**

vendor /ˈvendə(r)/ *n.* （尤指街上的）小贩 **U5**

versatile /ˈvɜːsətaɪl/ *adj.* 通用的；万能的 **U4**

vertically /ˈvɜːtɪkli/ *adv.* 垂直地 **U3**

veteran /ˈvetərən/ *adj.* 经验丰富的 **U3**

volatilization /ˌvɒlətɪlɪˈzeɪʃn/ *n.* 蒸发 **U8**

wander /ˈwɒndə(r)/ *v.* 漫步 **U9**

wasabi /wəˈsɑːbi/ *n.* 绿芥末 **U1**

Water Margin 《水浒传》 **U9**

wedding ceremony 婚礼 **U2**

whack /wæk/ *n.* 重击 **U4**

whetstone /ˈwetstəʊn/ *n.* 磨刀石 **U4**

witness /ˈwɪtnəs/ *v.* 目击；见证 *n.* 证人 **U2**

wok /wɒk/ *n.* 炒锅 **U1**

wok scoop 锅铲 **U4**

worship /ˈwɜːʃɪp/ *v.* 崇拜；敬奉 **U2**